Benedito Ferreira Bonifacio
Reginaldo Gomes Nobre
Evandro M. da Silva

Irrigation with saline water and potassium in guava rootstock

Benedito Ferreira Bonifacio
Reginaldo Gomes Nobre
Evandro M. da Silva

Irrigation with saline water and potassium in guava rootstock

Aspects of rootstock growth and quality

ScienciaScripts

This book is a translation from the original published under ISBN 978-613-9-61826-2.

Publisher:
Sciencia Scripts
is a trademark of
Dodo Books Indian Ocean Ltd. and OmniScriptum S.R.L publishing group

120 High Road, East Finchley, London, N2 9ED, United Kingdom
Str. Armeneasca 28/1, office 1, Chisinau MD-2012, Republic of Moldova, Europe
Printed at: see last page
ISBN: 978-620-7-70072-1

SUMMARY

GENERAL SUMMARY

In view of the socio-economic importance of guava growing, especially in the northeastern region of Brazil where there is limited availability of good quality water, there is a need to use saline water for agricultural production, implying the need to develop strategies that can make its use viable. With this in mind, the research was carried out with the aim of evaluating the effects of different doses of potassium combined with water of different saline levels on the production of seedlings for guava rootstocks. The experiment was carried out in a greenhouse at the Center for Agrifood Sciences and Technology of the Federal University of Campina Grande, Pombal - PB. The experimental design was in randomized blocks, in a 5 x 4 factorial scheme, with the treatments referring to five levels of electrical conductivity of the irrigation water (ECa = 0.3; 1.1; 1.9; 2.7 and 3.5 dS m^{-1}) in interaction with four doses of potassium (70, 100, 130 and 160% K) with the recommended dose of 100% K (726 mg of K dm^{-3} of substrate) for guava seedlings and four replications, each plot consisting of two useful plants. The treatments were applied at 40 days after seedling emergence (DAE). The rootstocks were evaluated at 120 and 225 DAE, using plant height, stem diameter, number of leaves and leaf area; from 60 to 225 DAE, the absolute and relative growth rates for plant height were measured and, at 225 DAE, the specific leaf area, the variables of dry phytomass of stem, leaves, root dry, aerial part, total, root/ aerial part ratio, leaf area ratio and the Dickson quality index. Irrigation with water with an ECa of up to 1.9 dS m^{-1} enabled the formation of guava rootstock cv. Paluma with an acceptable reduction in its growth; a potassium dose of 798.6 mg of K dm^{-3} of substrate promoted the greatest growth in height of the guava rootstock cv. Paluma at 120 days after emergence; increasing doses of K did not attenuate the harmful effects of salts on guava rootstocks cv. Paluma; a dose of 508.6 mg of K dm of substrate promoted the greatest growth in height of the guava rootstock cv. Paluma at 120 days after emergence; increasing doses of K did not attenuate the harmful effects of salts on guava rootstocks cv. Paluma. Paluma; a dose of 508.2 mg of K dm^{-3} of substrate favors the accumulation of dry phytomass of Paluma guava stems at 225 DAE; irrigation with water of ECa 1.9 dS m$^-$ 1 promotes an acceptable 10% reduction in phytomass production and quality of Paluma guava rootstocks; there was no significant interaction (salt x doses of K) on the variables studied.

Keywords: *Psidium guajava* L., salt stress, mineral nutrition.

GENERAL ABSTRACT

Considering the socioeconomic importance of the guava crop, especially for the northeastern region of Brazil where there is limited availability of good quality water, the need arises for the use of saline waters for agricultural production, implying the need to develop strategies that can their use. In this sense, the research was carried out to evaluate the effects of different doses of potassium combined with waters of different salt levels in the production of seedlings for guava rootstock. The experiment was carried out in a greenhouse at the Agrifood Science and Technology Center of the Federal University of Campina Grande, Pombal - PB. The experimental design was a randomized complete block design, in a 5 x 4 factorial scheme, with the treatments referring to five levels of electrical conductivity of the irrigation water (CEw = 0.3; 1.1; 1.9; 2.7 and 3.5 dS m^{-1}) in interaction with four doses of potassium (70, 100, 130 and 160% K) and the recommended dose of 100% K (726 mg K dm^{-3} substrate) for guava seedlings and four repetitions, each plot consisting oftwo useful plants. The treatments were started at 40 days after emergence of the seedlings (DAE). The rootstocks were evaluated at 120 and 225 DAE, through plant height, stem diameter, number of leaves and leaf area; in the period from 60 to 225 DAE, absolute and relative growth rates were measured for plant height and, at 225 DAE, the specific leaf area, stem dry matter, leaves, root dry matter, shoot, total root/shoot ratio, leaf area ratio and Dickson quality index. Irrigation with CEw water of up to 1.9 dS m^{-1} allowed the formation of a guava rootstock cv. Paluma with acceptable reduction in its growth; potassium dose of 798.6 mg of K dm^{-3} substrate promoted the highest growth in height of the guava rootstock cv. Paluma at 120 days after the emergency; increasing doses of K did not attenuate the harmful effects of the salts on guava rootstocks cv. Paluma; dose of 508.2 mg of K dm^{-3} of substrate favors the accumulation of dry phytomass of cv. Paluma at 225 DAE; irrigation with CEw water 1.9 dS m^{-1} promotes a 10% acceptable reduction on the phytomass production and quality of guava rootstocks cv. Paluma; there was no significant interaction (salt x doses of K) on the studied variables.

Keywords: *Psidium guajava* L., salt stress, mineral nutrition. ix

1. INTRODUCTION

The guava tree (*Psidium guajava* L.) is very important in the context of Brazilian fruit growing (CERQUEIRA et al., 2011) and is well accepted due to its pleasant taste and aroma, which are basic requirements both for *fresh* consumption and in the form of concentrated juices, jams, jellies, liqueurs, ice cream or other industrialized forms (CAVALCANTE et al., 2005a).

Worldwide, the main guava producing countries are India, Pakistan and Brazil, where it is grown in all geographical regions. The national production of guava in 2015 was 424.30 thousand tons in a harvested area of 17.68 thousand hectares, with an average yield of 24.10 tons per hectare. The Northeast and Southeast are the most productive regions, accounting for 48.82% and 42.25%, respectively, with the states of Pernambuco, Sao Paulo, Bahia, Rio de Janeiro and Minas Gerais being the main producing states (AGRIANUAL, 2013; IBGE, 2016).

There is a trend towards expanding the area under cultivation of this fruit tree, driven by the use of irrigation in the semi-arid areas of Brazil (CAVALCANTE et al., 2008a). However, the Brazilian semi-arid region is characterized by prolonged periods of drought, where there is a water deficit for plants due to the rate of evapotranspiration exceeding that of precipitation for most of the year (HOLANDA et al., 2010; CAVALCANTE et al., 2010). In this region, irrigation is essential to guarantee crop productivity, due to the occurrence of water insufficiency involving quantitative and qualitative aspects, especially with regard to the presence of salts in the water resources resulting from the climatic and geological characteristics of the region (MEDEIROS et al., 2003).

The scarcity of water for agriculture, coupled with the need to increase food production, requires the use of lower quality water. This has already been seen in several countries where producers have started to use brackish water for crop irrigation (SAVVAS et al., 2007). However, success will depend on the use of tolerant species and the adoption of appropriate crop, soil, irrigation water and fertilizer management practices (AYERS & WESTCOT, 1999; RHOADES et al.,

2000).

The effects of salinity on plants can be caused by the difficulty plants have in absorbing water, the toxicity of specific ions and interference with nutrient absorption (DIAS et al., 2005). These problems lead to loss of seedling quality, inhibition of growth, phytomass and production of cultivated plants (SOUZA et al., 2016).

The guava tree is classified as moderately sensitive to soil and water salinity, suffering a reduction in its productive potential when the electrical conductivity of the irrigation water exceeds 3.0 dS m^{-1} (AYERS & WESTCOT, 1999; CAVALCANTE et al., 2005b). Tàvora et al. (2001) showed that guava seedlings are more sensitive to salts than in the other stages of growth. According to the authors, the crop cannot tolerate electrical conductivity of the soil saturation extract (ECes) and irrigation water (ECa) greater than 1.2 dS m^{-1} and 0.8 dS m^{-1} , respectively.

Among the alternatives used to minimize the damaging effects of salts on plants, particularly in fruit trees, is the use of rootstock associated with the practice of fertilization (ABRANTES, 2015; SOUZA et al., 2016), making it possible to use saline water during the formation of seedlings and plant growth (CAVALCANTE et al., 2005b). In this sense, Cavalcante et al. (2010) and Silva et al. (2015) found that fertilizing with bovine biofertilizer and nitrogen, respectively, had a mitigating effect on the salinity of irrigation water in the initial growth of the "Paluma" guava tree; however, further studies with other nutrient sources, such as potassium, are needed to consolidate the recommendations on the use of lower quality water in the production of fruit tree seedlings in the semi-arid region.

Mineral nutrition is an important environmental factor, with potassium being one of the macronutrients most demanded by agricultural crops (NATALE et al., 1994; FRANCO et al., 2007) due to its functions in the physiological processes of plants, such as the activation of enzymes, the opening and closing of stomata, photosynthesis, as well as acting in the translocation of carbohydrates and protein synthesis (TAIZ & ZEIGER, 2013); neutralization of anions and the maintenance of osmotic potential (EPSTEIN & BLOOM, 2006). Studies show that supplementary application of KNO3 at a concentration of 5mM in fruit trees reduces the salt effect by increasing the K/Na, Ca/Na ratio and N absorption (KAYA et al., 2007).

2. OBJECTIVES

2.1 General

To evaluate the tolerance of guava rootstock cv. Paluma to irrigation water salinity, subjected to fertilization with different doses of potassium.

2.2 Specific objectives

• To evaluate the effect of irrigation water salinity on growth, phytomass and quality variables of guava rootstock cv. Paluma fertilized with different doses of potassium;

• To determine the level of irrigation water salinity tolerated by the Paluma guava tree for the production of viable rootstocks under different doses of potassium;

• To determine the appropriate dose of potassium for the production of guava rootstock cv. Paluma under irrigation with salinized water.

3. LITERATURE REVIEW

3.1 Guava growing

3.1.1 Botany and morphology

The guava tree belongs to the Mirtaceae family, which consists of 102 genera and 3024 species, distributed in tropical and subtropical climates. It is a small to medium-sized tree, usually 3 to 5 m high, and can reach 9 m in height, however, with pruning, they reach approximately 2.5 to 4 m in height (MANICA et al., 2001).

It has a pivoting root system, where lateral roots appear and branch out exuberantly near the soil surface, with a large number of root cells concentrated between 0 and 30 cm deep (MANICA et al., 2001). When propagated by rooting cuttings, they only have secondary roots and are shallower than plants propagated by seeds (GONZAGA NETO, 2007).

The leaves are opposite, elliptical-oblong in shape and fall off after ripening. The flowers are white, hermaphrodite and appear in single buds or in groups of two or three buds, always in the axil of the leaves that sprout on mature branches, after pruning or naturally (GONZAGA NETO, 2007).

The fruit of the guava tree is a berry with a mesocarp of varying thickness, firm texture and numerous seeds (MANICA et al., 2000). The seeds are reniform in shape, hard, between 2 and 3 mm in size and vary in number depending on the cultivar (ZAMBÂO & BELLINTANI, 1998; MANICA et al., 2001).

3.1.2 Origin and socio-economic aspects

Despite differences of opinion, it is believed that the guava tree originated in Tropical America, possibly between Mexico and Peru, where it can still be found in the wild (NATALE, 2009). Brazil is the world's third largest producer of guava, with the Northeast and Southeast being the most productive regions, accounting for approximately 91% of Brazilian production (IBGE, 2016).

In the Northeast, guava is grown in the states of Pernambuco, Bahia and Cearà (IBGE, 2016), where it is of great economic and social importance, especially as it is grown on properties that vary in size from three to five hectares, with a predominantly family workforce (NATALE et al., 1996).

Even though they occupy relatively small areas, orchards generate twenty times more produce per hectare than cereal crops, rewarding farmers and, as a rule, employing four more people per hectare than traditional crops, which helps to keep people in the countryside and reduce the rural exodus (NATALE, 2009).

3.1.3 Climate and soils

The climate range favorable to this crop is close to the equator, in low altitude locations, with average temperatures between 24 and 28°C, average relative humidity between 37 and 96% and rainfall of around 1,000 mm/year. Some of the limiting factors for good production can be droughts, poorly distributed rainfall or lack of irrigation (MANICA et al., 2000).

It grows well in deep, permeable, well-drained soils with a pH ranging from 4.5 to 8.0, adapting best to a pH range of 5.0 to 6.5. Crops planted in soils with a pH value of more than 7.0 have shown symptoms of iron deficiency (MANICA, 2000; PIO et al., 2014), with a base saturation (V%) of 70% and a minimum magnesium (Mg) content in the soil of 0.9 cmolc dm^{-3} (NATALE et al., 1996).

3.1.4 Paluma cultivar

Among the most widely planted cultivars in Brazil, the Paluma cultivar stands out for its exceptional characteristics for industrial processing and *fresh* consumption, as well as its productivity and vigor (PEREIRA et al., 2003; PEREIRA & NACHTIGAL, 2009).

The Paluma cultivar is a genetic material derived from the Rubi-Supreme guava, obtained from open-pollinated seeds (WATLINGTON, 2006). They are vigorous, lateral-growing and very fertile plants, and it is common for up to seventeen flower buds to appear on a single branch, which requires, after pruning, the practice of thinning the flowers *in* order to obtain fruit of the size and weight preferred by the consumer of *fresh fruit,* which, when thinned, can reach more than 500 g *in* the first productions; another quality of the Paluma is the post-harvest resistance of the fruit (GONZAGA NETO, 2007).

In terms of the fruit's chemical attributes, Paluma guava has high levels of total soluble solids (TSS) equivalent to 10° Brix, which gives the fruit an attractive flavor (LIMA et al., 2002; RAMOS et al., 2010) and is therefore the most widely planted cultivar in Brazil. With regard to total titratable acidity (TTA), Lima et al. (2002) found 0.63% citric acid in Paluma guava. Another important characteristic is the vitamin C content of the fruit, which was 89.78 mg of ascorbic acid/100 g.

3.1.5 Plant propagation

The production of guava seedlings is one of the main stages in obtaining early, healthy plants with high production potential. In this process, the genetic material, the propagation method, the quantity and quality of water available to the farmer for irrigating the seedlings are essential (QUEIROZ et al., 2010).

In fruit growing, propagation by seed is mainly used to obtain rootstocks, also known as "horses" in species such as citrus, avocado and guava, among others. However, the use of seeds to obtain seedlings for a commercial orchard has some limitations, such as juvenility, high plant vigor and high genetic variability, even between plants originating from the same matrix (FACHINELLO et al., 2005).

The use of seeds to propagate guava trees causes great variation, especially in terms of shape, growth habit, plant height, productivity and fruit characteristics, which is an obstacle to the commercial value of the crop (DIAS et al., 2014).

Vegetative, asexual or clonal propagation is the main form of seedling production in fruit growing (HARTMANN et al., 2002). It is extremely important in the formation of commercial orchards, as they have the same genetic makeup as the parent plant, with identical climatic, edaphic, nutritional and management needs (FACHINELLO et al., 2005).

Asexual propagation of the guava tree can be carried out by alporchia, grafting (budding or budding), cuttings (root or branch cuttings) and tissue culture (MANICA et al., 2001), although grafting and cuttings are the most widely used methods in fruit growing (HARTMANN et al., 2002).

Grafting is a technique that consists of joining parts of two or more plants which, through the regeneration of their tissues, become a single plant. In this sense, a grafted plant is made up of two parts, the rootstock, whose tissues constitute the root system, the basis for support and absorption of water and nutrients from the soil for the new plant, and the scion, whose tissues will constitute the crown of the plant (NACHTIGAL et al., 2005).

Rootstocks can confer important characteristics to plants, such as greater or lesser vigor to the scion, better fruit quality, tolerance to unfavorable conditions such as clay soils, excess or lack of humidity, pest and disease attacks (NACHTIGAL et al., 2005), tolerance to irrigation water salinity in fruit trees such as citrus (BRITO et al., 2014), cashew (SOUSA et al., 2011) and guava (TAVORA et al., 2001; SILVA et al., 2015; ABRANTES, 2015; SOUZA et al., 2016; SA et al., 2016).

Grafting on guava trees should be by slit grafting at the top or side, on rootstocks with a single, erect stem and a minimum of eight mature green leaves. The height should be between 15 and 25 cm, the diameter should be between 0.40 and 0.50 cm in the region of the scion and it should be free of pests, diseases and weeds; the forks should have the same diameter as the rootstock (CHAVES et al., 2000).

According to Manica et al. (2001), in the production of commercial guava seedlings by cuttings, herbaceous cuttings are the most commonly used. They should be selected from mother plants and taken from branches of the last vegetative flush, which are not yet lignified and green in color.

3.2 Water quality for irrigation

In order to correctly interpret the quality of irrigation water, the parameters analyzed must be related to their effects on the soil, the crop and irrigation management, which will be necessary to control or compensate for problems related to water quality (BERNARDO et al., 2006).

According to Ayers & Westcot (1999), irrigation water is classified according to three parameters: the first refers to the risk of salinity, which occurs when there is an accumulation of salts in the root zone at a certain concentration, causing a drop in yield; the second parameter refers to the risk of sodicity or infiltration problems, which occurs when relatively high levels of sodium or low levels of calcium and magnesium in the soil and water, through the dispersing action of sodium in soil colloids, reduce hydraulic conductivity; and the third concerns toxicity from specific ions, such as sodium, chloride and boron, which accumulate in plants in high concentrations, reducing the yield of crops sensitive to these ions.

As far as their chemical composition is concerned, water in nature, whatever its source, is made up in greater proportions of sodium, calcium, magnesium and potassium salts, in the form of chlorides, sulphates, bicarbonates and carbonates, although the quantity and type of these salts can vary greatly depending on the source, geographical location and time of collection (RICHARDS, 1954; MEDEIROS et al., 2003). The following are established as chemical indicators of water quality: pH (hydrogenionic potential), alkalinity, hardness, chlorides, iron and manganese, nitrogen, phosphorus and others (FONTES et al., 2011).

According to Maas (1990), water quality is only one of the factors that determine the type and intensity of management practices for the safe use of saline water in irrigation; other factors such as crop tolerance to salinity, soil properties, climate and crop management must be considered.

In Brazil, the classification of water suitable for irrigation is that proposed by the University of California Committee of Consultants (1974), and that of the United States Salinity Laboratory (RICHARDS, 1954). For classification in terms of seepage risk and ionic toxicity, the guidelines presented by Ayers & Westcot (1999) are more appropriate. Rhoades et al. (2000), propose the classification of saline waters according to electrical conductivity (EC) in dS m^{-1} and salt concentration in mg L^{-1} .

3.3 Use of saline water in the production of guava seedlings

In the arid and semi-arid regions of northeastern Brazil, where low rainfall and erratic rainfall predominate, the concentration of salts in the water is increased during the dry season (CAVALCANTE et al., 2008a), resulting in adversity to plants. In this region, due to inadequate irrigation management and the existence of a negative water balance, which limits the leaching of salts from the soil and contributes to their accumulation, there is salinization of irrigated areas (NOBRE et al., 2011).

In view of the low supply of good quality water for irrigation, the adoption of research into the feasibility of using saline water to grow food and non-food crops should be encouraged (PAULUS et al., 2010; SANTOS et al., 2013). In this sense, according to Savvas et al. (2007), in various parts of the world farmers have been successfully using brackish water for crop irrigation, however, the results depend on the use of tolerant species, phenological stage, salt characteristics, intensity and duration of saline stress, crop and irrigation management and soil and climatic conditions (ASHRAF & HARRIS, 2004).

As for the effect of salts, the guava crop is moderately sensitive to soil and water salinity, suffering a decline in its productive capacity in places where the electrical conductivity of the irrigation water exceeds 3.0 dS m^{-1} (AYERS & WESTCOT, 1999; CAVALCANTE et al., 2005b).

Souza et al. (2016) evaluated the effects of using different doses of nitrogen combined with water of different saline levels on the production of guava rootstock, finding greater growth and quality for the 'Crioula' guava rootstock when using water with an electrical conductivity of 0.3 dS m^{-1} and fertilizing with 541.1 mg of N dm^{-3} of soil; increasing doses of N do not reduce the deleterious effect of irrigation with salinized water on the growth of 'Crioula' guava rootstock; irrigation with water with an ECa of up to 1.75 dS m^{-1} , in the production of guava rootstock promotes an acceptable reduction in the growth and quality of seedlings of 10%, according to (AYERS & WESTCOT, 1999).

Abrantes (2015), when studying the effect of different doses of nitrogen on the production of grafted guava seedlings irrigated with water of different saline levels in the semi-arid northeast of Brazil, found that increasing the ECa from 0.3 dS m^{-1}

negatively affected growth variables and the quality of Paluma guava seedlings grafted onto Paluma and Crioula rootstocks. However, this author pointed out that seedlings irrigated with water with an ECa of up to 1.9 dS m^{-1} met the criteria for producing standard guava seedlings and the use of water with an ECa of up to 2.2 and 2.0 dS m^{-1} in the irrigation of Paluma guava seedlings, with the Crioula and Paluma rootstocks, respectively, promoted an acceptable 10% reduction in plant growth.

SA et al. (2016), when studying the tolerance of guava rootstocks under saline stress, found that increasing salinity restricts emergence, growth and phytomass accumulation, and the most drastic effects occur at levels above 1.8 dS m^{-1} ; the cultivar 'Crioula' is more tolerant to salinity than 'Paluma' and 'Ogawa', and can be indicated as a rootstock.

Silva et al. (2015) when assessing the initial growth of Paluma guava rootstocks irrigated with salinized water and nitrogen fertilization, found that increasing the ECa from 0.3 dS m^{-1} negatively affects rootstock growth, especially at 80 DAE (days after emergence) and that nitrogen fertilization at a dose of 541 mg dm^{-3} reduces the effect of irrigation water salinity on the stem diameter of Paluma guava seedlings. Paluma. Stem diameter is an important characteristic, since the greater its value, the greater the vigor, robustness and resistance of the plant (GUIMARAES et al., 2009).

Cavalcante et al. (2010) found that Paluma guava seedlings irrigated with saline water increased the salinity of the soil, which affected growth in height, stem diameter, leaf area, guava biomass production and the physiological and metabolic processes of the crop.

Gurgel et al. (2007) found, when studying the effect of irrigation water salinity on the initial growth of 'Rica' and 'Ogawa' guava rootstocks, that an increase in the electrical conductivity of the irrigation water caused deleterious effects on the growth of the cultivars in terms of number of leaves, leaf area, stem diameter, plant height and dry phytomass, more markedly in the Ogawa cultivar, and in the cv. Rica, the organ most affected was the root.

3.4 Effects of salinity on plants

3.4.1 Osmotic, toxic and indirect effects

The osmotic effect is one of the main problems that salinity exerts on plants, due to the large presence of salts in the soil solution reducing the availability of water as a result of the decrease in osmotic potential in the root zone (MOURA, 2000).

The saltier the water, the harder it is to extract it from the soil solution. Depending on the degree of salinity, the plant may lose the water inside the roots instead of absorbing it, due to the loss of water from the cells to the concentrated solution in the soil through a process known as plasmolysis (DIAS & BLANCO, 2010).

According to Dias et al. (2003), crops are not equally affected by salinity levels, as some are more tolerant than others and can extract water more easily. The response of crops to saline stress varies between species and cultivars and, for the same species and cultivar, with the phenological stage of the crop, environmental conditions, soil and irrigation water management (MAAS, 1990).

Each plant species or cultivar can tolerate up to a certain salinity without reducing potential yield, also known as threshold salinity (SL), after which there is a reduction in productivity as soil salinity increases. Some produce economically acceptable yields under high salinity levels, while others are sensitive to relatively low levels. This can be explained by the better osmotic adaptation capacity of some plants, which allows them to absorb sufficient amounts of water even in saline environments (AYERS & WESTCOT, 1999).

Toxicity problems usually arise when certain ions, constituents of the soil or water, are absorbed by plants and accumulate in their tissues in concentrations high enough to cause damage and reduce yields (AYERS & WESTCOT, 1999). According to Pizarro (1985), the toxic effect of specific ions is not due to the direct effect of the ions, but because they induce changes in metabolism, causing the accumulation of toxic products.

In plants growing in soils affected by salts, for example, there is a decrease in the concentration of K^+ in the cell cytosol caused by a greater increase in the concentration of Na^+ (ZHU, 2003). This lower absorption of K^+ has been attributed to greater competition between Na^+ and K^+ for adsorption sites or to a greater efflux of K^+ from the roots (MARSCHNER, 2005).

The main consequences of the accumulation of Na^+ and Cl^- ions in the leaves include necrosis of leaf tissues and accelerated senescence of mature leaves, thus reducing the area set aside for photosynthesis (MUNNS et al., 2006). Thus, photosynthetic activity is limited, not only by the stomatal closure caused by osmotic stress, but also by the effect of salts on the leaves.

According to Silva (2011), chlorine and sodium ions, besides being the most present in irrigation water, can be absorbed by the roots, translocated and accumulated in the leaves, or directly by the wet leaves during sprinkler irrigation, especially during periods of high temperatures and low humidity.

The increase in salts in irrigation water causes losses in growth and production in most plants, due to the toxic effect of excess Na^+ and Cl- ions, which can cause a reduction in water and nutrient absorption and an imbalance in the cation balance and plant metabolism (MARSCHNER, 2005; MUNNS & TESTER, 2008; NIVAS et al., 2011).

The indirect effects caused by salinity occur when high concentrations of sodium or other cations in the solution interfere with the physical conditions of the soil or the availability of certain elements, indirectly affecting plant growth and development (DIAS et al., 2003). According to Tester & Davenport (2003), the presence of certain ions in excess can prevent the absorption of other essential elements for plant growth, leading to nutritional imbalance.

3.5 Potassium fertilization

The guava tree is considered a rustic plant, adaptable to various types of soil, however, substantial increases in the production of this fruit can be achieved if proper fertilizer management is used (NATALE, 2003). It has been observed that fruit tree nutrition influences the size, weight, external appearance of the fruit, post-harvest quality and preservation, resistance to pests and diseases, among other aspects (NATALE & MARCHAL, 2002).

According to Natale et al. (1994), the macronutrients most exported by guava fruit are: K, N, P, S, Mg and Ca, while the micronutrients are extracted in the following ascending order: B, Cu, Zn, Fe and Mn. Franco et al. (2007) determined, using hydroponics, the nutrient absorption rate in guava seedlings, with nitrogen and potassium being the two elements most required by the crop at this stage of development.

Plants need large amounts of potassium, as it is removed by most crops more than any other nutrient, indicating the need to apply an adequate amount of potassium fertilizers (DEV, 1995). Potassium occupies a prominent place, given the deficiency of this nutrient in most soils in the Northeast, and because it is a nutrient with various roles in plant metabolism (MALAVOLTA, 2005).

Potassium is a nutrient with major functions in the physiological processes of plants, through the activation of enzymes, in the opening and closing of stomata, regulating tissue turgidity, controlling the concentration of CO_2 in the sub-stomatal chamber, which is essential for photosynthesis, resulting in biomass production; It also acts in the translocation of carbohydrates and protein synthesis (TAIZ & ZEIGER, 2013); it neutralizes anions and regulates osmotic potential (EPSTEIN & BLOOM, 2006).

Potassium deficiency normally reduces the size of internodes, apical dominance and plant growth, delays fruiting and produces smaller fruit with less color intensity (ERNANI et al., 2007). However, the high availability of K favors consumption, which deserves attention because it interferes with the absorption of other nutrients, such as Ca^{2+} and Mg^{2+} (BATAGLIA, 2005).

4. BIBLIOGRAPHICAL REFERENCES

ABRANTES, D. S. **Interaction between salinized water and nitrogen fertilization in the production of grafted guava seedlings.** 2015. 91 f. Dissertation (Master's Degree in Agroindustrial Systems) - Federal University of Campina Grande, Pombal, 2015.

AGRIANUAL 2013: Brazilian production yearbook. Sao Paulo: **FNP Consultoria e Comércio**, 334p. 2013.

AYERS, R. S.; WESTCOT, D. W. **Water quality in agriculture**. Campina Grande: Federal University of Paraiba, 1999. 153p.

ASHRAF M.; HARRIS, P.J.C. Potential biochemical indicators of salinity tolerance in plants.**Plant Science**, Chicago, v.166, n.1, p.3-16, 2004.

BATAGLIA, O.C. Diagnostic methods for potassium nutrition with emphasis on DRIS. In: YAMADA, T.; ROBERTS, T.L (eds). Potassium in Brazilian agriculture. SYMPOSIUM ON POTASSIUM IN BRAZILIAN AGRICULTURE, São Paulo, SP, 2004. **Proceedings...** Piracicaba: POTAFOS, 2005, chap.13, p.322-241.

BERNARDO, S.; SOARES, A. A.; MANTOVANI, E. C. **Manual de irrigaçâo**. 8 ed., Viçosa: UFV, 2006. 625 p.

BRITO, M. E. B.; FERNANDES, P. D.; GHEYI, H. R.; MELO DE, A. S.; SOARES FILHO, W. dos S.; SANTOS DOS, R. T. Sensitivity to salinity of trifoliate hybrids and other citrus rootstocks. **Revista Caatinga**, Mossoró, v.27, n.1, p.17-27, 2014.

CAVALCANTE, I. H. L.; CAVALCANTE, L. F.; OLIVEIRA de, F. A.; ARAÙJO de, F. A. R. Production, nutrient export and mineral composition in two guava genotypes. **Cientifica**, Jaboticabal, v. 33, n. 2, p. 112-119, 2005a.

CAVALCANTE, L. F.; CAVALCANTE, I. H. L.; PEREIRA, K S. N.; DE OLIVEIRA, F. A.; GONDIM, S C.; DE ARAÙJO, F A. R. Germination and initial growth of guava plants irrigated with saline water. **Revista Brasileira de Engenharia Agricola e Ambiental**, Campina Grande, v. 9, n. 4, p. 515-519, 2005b.

CAVALCANTE, I. H. L.; SILVA, G. F.; CAVALCANTE, L. F.; SANTOS, D.; BECKMANCAVALCANTE, M. Z. Mineral composition of Paluma guava leaves as a function of sulfate-nitrogen fertilization. **Revista Brasileira de Ciências Agrârias**, Recife, v. 3, n.1, p. 6-12, 2008a.

cAVALcANTE, L. F.; SILVA, M. N. B.; DINIZ, A. A.; cAVALcANTE, I. H. L.; CAMPOS, V. B. Biomass of yellow passion fruit in soil irrigated with saline water protected against water loss. **Revista Verde de Agroecologia e Desenvolvimento Sustentâvel**, Mossoró, v.3, n.3, p.26-34, 2008b.

cAVALcANTE, L. F; VIEIRA, M. S.; SANTOS, A. F.; OLIVEIRA, W. M.; NASCIMENTO, J. A. M. Saline water and liquid bovine manure in the formation of seedlings of guava cultivar Paluma. **Revista Brasileira de Fruticultura**, Jaboticabal, v. 32, n. 01, p. 251-261, 2010.

cERQUEIRA, T.S.; JAcOMINO, A.P.; SASAKI, F.F.; ALLEONI, A.c.c. **Coating guavas with protein and chitosan films**. BRAGANTIA. campinas, v. 70, n. 1, p.216-221, 2011.

CHAVES, J. C. M. CAVALCANTI JÛNIOR, A. T.; CORREIA, D.; SOUZA, F. X.; ARAÙJO, C. A. T. **Norms for seedling production**. Fortaleza: Embrapa Agroindùstria Tropical, 2000. 37p. (Documents, 41).

DEV, G. Potassium - An essential nutrient. In: Use of Potassium in Punjab Agriculture. **Potash and Phosphate Institute of Canada**, Gurgaon: India Programme, 1995.

DIAS, N.S.; GHEYI, H.R.; DUARTE, S.N. **Prevention, management and recovery of soils affected by salts**. Piracicaba: ESALQ, Department of Rural Engineering, 2003. 118p. Didactic Series, 13.

DIAS, N. da S.; DUARTE, S. N.; TELES FILHO, J. F.; YOSHINAGA, R. T. Lettuce production under different levels of soil salinity. **Irriga**, Botucatu, v.10, p.20-29, 2005.

DIAS, N. S.; BLANCO, F.F. Effects of salts on soil and plant. **In:** GHEYI, H. R.;

DIAS, N. da S.; LACERDA, C. F. (Org.). Management of salinity in agriculture: basic and applied studies. Fortaleza: National Institute of Science and Technology in Salinity, 2010, p. 133-144.

DIAS, J. M. M.; FELISMINO, D. C.; MOTOIKE, S. Y.; SIQUEIRA, D. L.; BRUCKNER, C. H. **Propagation of guava trees**. 2014. Available at: http://www.nutricaodeplantas.agr.br/site/ensino/pos/Palestras_William/Livrogoiaba_ pdf/16_propagacaogoiaba.pdf. Accessed on: March 16, 2016.

EPSTEIN, E.; BLOOM, A. J. **Nutriçâo mineral de plantas:** principios e perspectivas. 2. ed. Londrina: Editora Planta, 2006. p. 169-236.

ERNANI, P. R.; ALMEIDA, J. A.; SANTOS, F. C. Potassium. In: NOVAIS, R. F.; ALVAREZ V., V. H.; BARROS, N. F.; FONTES, R. L. F.; CANTARUTTI, R. B.; NEVES, J. C. L. **Fertilidade do solo.** Viçosa: SBCS, 2007. chap. 9, p. 551-594.

FACHINELLO, J. C.; NACHTIGAL, J.C.; HOFFMANN, A. Seed propagation. IN: FACHINELLO, J. C.; HOFFMANN, A.; NACHTIGAL, J.C. (Ed.). **Propagation of fruit plants.** Brasilia, DF: Embrapa Informaçâo Tecnològica. 2005. P. 57-67.

FONTES, R. L. F. F.; CANTARUTTI, R. B.; NEVES, J. C. L. Fertilidade de Solo. **Brazilian Society of Soil Science**, Viçosa, 2011.

FRANCO, F. C.; PRADO, R. M.; BRACHIROLLI, L. F.; ROZANE, D. E. Growth curve and macronutrient uptake in guava seedlings. **Revista Brasileira de Ciência do Solo**, Viçosa, v. 31, n. 6, p. 1429-1437, 2007.

GONZAGA NETO, L. **Guava production**. 14^a Semana Internacional da Fruticultura, Floricultura e Agroindùstria - FRUTAL September 10 to 13, 2007 - Centro de Convençôes do Cearà Fortaleza - Cearà - Brazil.

GUIMARAES, A. S.; MACEDO, B. N. E.; COSTA, S. G. Increasing sources and doses of organic and mineral fertilizers in the initial growth of jatropha. **Mens agitat**, Boa Vista, v. 04, n. 1, p. 17-22, 2009.

GURGEL, M. T.; GHEYI, H. R.; FERNANDES, P. D.; SANTOS, F. J. S.; NOBRE,

R. G. Initial growth of guava rootstocks irrigated with saline water. **Revista Caatinga**, Mossoró, v.20, n.2, p.24-31, 2007.

HARTMANN, H. T.; KESTER, D. E.; DAVIES JÙNIOR, F. T.; GENEVE, R. L. **Plant propagation:** principles and practices. 7th ed. New Jersey: Prentice Hall, 2002. 880 p.

HOLANDA, J. S.; AMORIM, J. R. A.; FRRREIRA NETO, M.; HOLANDA, A. C. Water quality for irrigation. In: GHEYI, H. R.; DIAS, N. S.; LACERDA, C. F (ed). **Salinity management in agriculture**: Basic and applied studies. FORTALEZA, INCTA Sal, 2010. p. 43-61.

IBGE - Brazilian Institute of Geography and Statistics (2016). Sidra - **Municipal Agricultural Production**, 2016. Available at:< http:// www.sidra.ibge.gov.br> (Accessed December 12, 2016).

KAYA, C.; TUNA, A. L.; ASHRAF, M.; ALTUNLU, H. Improved salt tolerance of melon (*Cucumismelo L.*) by the addition of proline and potassium nitrate. **Environmental and Experimental Botany**, v. 60, p. 397-403, 2007.

LIMA, M. A. C. DE; ASSIS, J. S. DE; GONZAGA NETO, L. Characterization of guava fruit and selection of cultivars in the Submedio Sao Francisco region. **Revista Brasileira de Fruticultura,** Jaboticabal, v.24, n.1, p.273-276, 2002.

MAAS, E. V. Crop salt tolerance. In: TANJI, K. K. **Agricultural salinity assessment and management**. New York: ASCE, 1990. ch. 13, p. 262-304.

MALAVOLTA, E. Potassium: absorption, transport and redistribution in the plant. In: YAMADA, T.; ROBERTS, T. L. (ed.). **Potassium in Brazilian agriculture**. Piracicaba: Brazilian Association for Potash and Phosphate Research, 2005. chap. 8, p. 179-238.

MANICA, I. Economic importance. In: MANICA, I.; ICUMA, I. M.; JUNQUEIRA, N. T. V.; SALVADOR, J. O.; MOREIRA, A.; MALAVOLTA, E. (eds.). **Tropical fruit growing 6**: guava. Cinco Continentes, Porto Alegre, p.9-22, 2000.

MANICA, I.; ICUMA, I. M.; JUNQUEIRA, N. T. V.; SALVADOR, J. O.;

MOREIRA, A.; MALAVOLTA, E. **Goiaba: From planting to consumer: Production technology, post-harvest, marketing.** Porto Alegre: CincoContinentes, 2001. 124 p.

MARSCHNER, H. **Mineral nutrition of higher plants.** 2 ed. London: Academic Press, 2005, 889 p.

MEDEIROS, J. F. DE; LISBOA, R. A.; OLIVEIRA, M.; SILVA JÚNIOR, M.J.; ALVES, L. P. Characterization of groundwater used for irrigation in the Chapada do Apodi melon-producing area. **Revista Brasileira Engenharia Agricola e Ambiental,** Campina Grande, v. 7, n.3, p. 469- 472, 2003.

MOURA, R. F. de. **Effects of soil recovery leaching rates and irrigation water salinity on the production components and crop coefficient of sugar beet.** Viçosa, 2000, 119p. Thesis (Doctorate in Agricultural Engineering) - Federal University of Viçosa, Viçosa, MG, 2000.

MUNNS, R.; JAMES, R.A.; LĂUCHLI, A. Approaches to increasing the salt tolerance of wheat and other cereals.**Journal of Experimental Botany,** Oxford, v.57, p.1025-1043, 2006.

MUNNS, R.; TESTER, M. Mechanisms of salinity tolerance.**Annual Review of Plant Biology,** v.59, n.1, p.651-681, 2008.

NACHTIGAL, J. C.; FACHINELLO, J. C.; HOFFMANN, A. Vegetative propagation by cuttings. IN: FACHINELLO, J. C.; HOFFMANN, A.; NACHTIGAL, J. C. (Ed.). **Propagation of fruit plants.** Brasilia, DF: Embrapa Informaçâo Tecnològica. 2005. P. 111-147.

NATALE, W. **Liming, fertilization and nutrition of the guava crop.** In: Guava cultivation: technology and the market. Edited by Danilo Eduardo Rozane, Flavio A. d' Araujo Couto, Empresa Junior de Agronomia. Viçosa: UFV; EJA, 2003, pp. 301-331. NATALE, W.; MARCHAL, J. Adsorption and redistribution of nitrogen (N) in citrus. **Revista Brasileira de Fruticultura,** Jaboticabal, v.24, n.1, p.183-188, 2002.

NATALE, W.; **Liming, fertilization and nutrition of the guava crop.** FCAV/Unesp, Campus Jaboticabal, 25p, 2009.

NATALE, W.; COUTINHO, E. L. M.; BOARETTO, A. E.; PEREIRA, F. M. **Guava: liming and fertilization**. Jaboticabal: FUNEP, 1996. 22 p.

NATALE, W., COUTINHO, E. L. M., BOARETTO, A. E., CORTEZ G. E. P., FESTUCCIA, A.J. **Extraction of nutrients by guava fruits (*Psidium guajava* L.)**. Cientifica, Sao Paulo, v.22, n.2, p.249-253, 1994.

NIVAS, D.; GAIKWAD, D.K.; CHAVN, P.D. Physiological responses of two Morinda Species under saline conditions. **American Journal of Plant Physiology**. p.1-10, 2011.

NOBRE, R. G.; GHEYI, H. R.; SOARES, F. A. L.; CARDOSO, J. A. F. Sunflower production under salt stress and nitrogen fertilization. **Revista Brasileira de Ciências do Solo**, Viçosa, v. 35, n. 3, p. 929-937, 2011.

PAULUS, D.; DOURADO NETO, D.; FRIZZONE, J. A.; SOARES. T. M. Production and physiological indicators of lettuce under hydroponics with saline water. **Horticultura Brasileira**, Brasilia, v.28, n.1, p.29-35, 2010.

PEREIRA, F. M.; NACHTIGAL, J. C. Genetic Improvement of Guava. *In Anais III.* **Simpósio Brasileiro da Cultura da Goiaba.** v. 2. Jaboticabal: Unesp, p. 371398, 2009.

PEREIRA, F.M.; CARVALHO, C.; NACHTIGAL, J.C. Século XXI: a new dual-purpose guava cultivar. **Revista Brasileira de Fruticultura**, Jaboticabal, v. 25, n.3, p.498-500, 2003.

PIO, R.; VALE, M. R.; JUNKEIRA, K. P.; RAMOS, J. D. **Guava tree culture**. 2014. Available at:<www.editora.ufla.br/index.php/component/.../56-boletins-de-extensao?>. Accessed: March 30, 2016

PIZARRO, F. **Drenage agricola y recuperacion de suelos salinos**. 2. Ed. Madrid: Editorial Espanola S.A. 1985. 542p.

QUEIROZ, J. E.; GONÇALVES, A. C. A.; SOUTO, J. S.; FOLEGATTI, M. V. Evaluation and monitoring of soil salinity. In: GHEYI, R. H.; DIAS, N. S.; LACERDA, C.F. **Management of salinity in agriculture**: Basic and applied studies. Fortaleza: INCT, 2010. 472 p.

RAMOS, D. P.; SILVA, A. C.; LEONEL, S.; COSTA, S. M.; DAMATTO JÚNIOR, E. R. Fruit production and quality of the 'Paluma' guava tree, subjected to different pruning times in a subtropical climate. **Revista Ceres**, Viçosa, v. 57, n. 5, p. 659-664, 2010.

RHOADES, J.; KANDIAH, A.; MASHALI, A. M. Use of aguasalinas for agricultural production. Translated by GHEYI, H. R., SOUSA, J. R. de, QUEIROZ, J. E. Campina Grande: UFPB. 2000. 117p. **(Estudos FAO Irrigaçao e Drenagem 48)**.

RICHARDS, L. A. (ed.) **Diagnosis and improvement of saline and alkali soils**. Washington, United States Salinity Laboratory: 1954. 160p. (USDA. Agriculture Hand book, 60).

SA, F. V. S; NOBRE, R. G.; SILVA, L. A.; MOREIRA, C. L.; PAIVA, E. P.; OLIVEIRA, F. A. Tolerance of guava rootstocks to salt stress. **Revista Brasileira de Engenharia Agricola e Ambiental**, Campina Grande, v.20, n.12, p.1072-1077, 2016.

SANTOS, J. B.; SANTOS. D. B.; AZEVEDO, C. A. V.; REBEQUI, A. M.; CAVACANTE, L. F.; CAVALCANTE, I. H. L. Morphophysiological behavior of the BRS Energia papaya tree subjected to irrigation with saline water. **Revista Brasileira de Engenharia Agricola e Ambiental**, Campina Grande, v.17, n.2, p.145-152. 2013. SAVVAS, D.; E. Stamati.; Tsirogiannis, I. L.; Mantzos N.; Barouchas, P.E.; Katsoulas N.; Kittas C. Interactions between salinity and irrigation frequency in greenhouse pepper grown in closed-cycle hydroponic systems. **Agricultural Water Management,** Amsterdam, v. 91, n. 1, p. 102-111, 2007.

SILVA, E. M.; NOBRE, R. G.; GHEYEI, H. R.; PINHEIROS, F. W. A.; SOUZA, L.

de P.; DIAS, A. S. **Growth and quality of guava rootstock under irrigation with salinized water and nitrogen doses.** In: III Inovagri International Meeting, Fortaleza, CE. 2015.

SILVA, I. N. Water quality in irrigation. **ACSA - Agricultura Cientifica no Semi Arido**, Patos v.07, n 03, julho - setembro, p. 01 - 15, 2011.

SOUSA, A. B. O.; BEZERRA, M. A.; FARIAS, F. C. Germination and initial development of common cashew clones under irrigation with saline water. **Revista Brasileira de Engenharia Agricola e Ambiental**, Campina Grande, v.15, p.390-394, 2011.

SOUSA, L. DE P., NOBRE R. G., SILVA E. M.; LIMA, de G. S.; PINHEIRO, F. W. A.; ALMEIDA L. L. de S. Formation of 'Crioula' guava rootstock under saline water irrigation and nitrogen doses. **Revista Brasileira de Engenharia Agricola e Ambiental,** Campina Grande, v.20, n.8, p.739-745. 2016.

TAIZ, L.; ZEIGER, E. **Plant physiology.** 5. ed. Porto Alegre: Artmed, 2013. 954p.

TAVORA, F. J. A. F.; PEREIRA, R. G. HERNADEZ, F. F. F. Growth and water relations in guava plants submitted to salt stress with Na Cl. **Revista Brasileira de fruticultura**, Jaboticabal, v. 23, n 2, p. 441- 446, 2001.

TESTER, M.; DAVENPORT, R. **Na^+ Tolerance and Na^+ transport in higher plants.** Annals of Botany, v.91, n.5, p.503-527, 2003.

WATLINGTON, F. Guava in the World. University of Puerto Rico, Rio Piedras, Puerto Rico. **The Agronomist.** Campinas, p. 58, 2006.

UNIVERSITY OF CALIFORNIA COMMITTEE OF CONSULTANTS. **Guidelines for interpretation of water quality for agriculture.** Davis: University of California, 1974. 13 p.

ZAMBÂO, J. C.; BELLINTANI NETO, A. M. **Guava Culture**. Campinas. CATI, 1998.23 p. (Technical bulletin CATI 236).

ZHU, J. K. Regulation of ion homeostasis under salt stress. **Current Opinion in**

Plant Biology, v. 06, n. 05, p. 441-445, 2003.

CHAPTER I GOIABEIRA 'PALUMA' SPROUT FORMATION IRRIGATED WITH SALINE WATER AND POTASSIC FERTILIZER

GOIABEIRA 'PALUMA' SPRING FORMATION

IRRIGATED WITH SALINE WATER AND POTASSIUM FERTILIZATION

ABSTRACT: Excess salts in soils and/or irrigation water affect the quality of seedlings. The aim of this research was to evaluate the effects of using salinized water associated with potassium fertilization on the growth of guava rootstock cv. Paluma. The experiment was conducted from March to December 2015, in a greenhouse at the UFCG Campus, Pombal, PB, using a randomized block design, in a 5 x 4 factorial scheme, with 4 replications, with the treatments consisting of five levels of electrical conductivity of the irrigation water (ECa = 0.3; 1.1; 1.9; 2.7 and 3.5 dS m^{-1}) and four doses of potassium (508.2; 726; 943.8 and 1.161.6 mg of K dm^{-3} of substrate), corresponding to (70, 100, 130 and 160% of K) with the recommended dose being 100% K (726 mg of K dm^{-3} of substrate). Irrigation with water with an ECa of up to 1.9 dS m^{-1} enabled the formation of guava rootstock cv. Paluma with an acceptable reduction in its growth; potassium doses of 798.6 mg of K dm^{-3} promoted the greatest growth in height of guava rootstock cv. Paluma at 120 days after emergence. Paluma rootstock at 120 days after emergence; increasing doses of K did not mitigate the harmful effects of salts on the growth variables evaluated; and the effects of irrigation water salinity and potassium fertilization on the growth of guava rootstocks are independent.

Keywords: *Psidium guajava* L., salinity, potassium.

1 INTRODUCTION

The semi-arid region of Brazil is characterized by climatic instability and water insufficiency, due to the region's high evaporation rate, so the production system depends on irrigation (ALMEIDA, 2010; HOLANDA et al., 2010). However, existing water sources have high salt concentrations (MEDEIROS et al., 2003), with values often reaching 5.0 dS m^{-1} (NEVES et al., 2009). Thus, the use of these waters is dependent on the tolerance of the crops to salinity and on irrigation and fertilization management practices, which should avoid environmental impacts and damage to the crops (LIMA et al., 2015).

Excess salts in the soil reduce the availability of water for plants, due to a reduction in the osmotic potential of the solution (SOARES et al., 2007), in addition to promoting specific toxicity and/or nutritional disorders, inducing morphological, structural and metabolic changes in higher plants (MEDEIROS et al., 2010), to the point of negatively affecting the quality of seedlings and plant production (BEZERRA et al., 2016). The guava tree is considered moderately tolerant to salinity, suffering a reduction in productivity when the electrical conductivity of the irrigation water exceeds 3.0 dS m^{-1} (CAVALCANTE et al., 2005).

Guava cultivation is a socio-economically viable crop in Brazilian agribusiness, helping to keep people in the countryside and improving regional income distribution (NUNES et al., 2014). It is widely cultivated in irrigated areas in the semi-arid region and is one of the most economically valuable fruit crops in the Northeast of Brazil (CAVALCANTE et al., 2010). Among the biological materials available, the Paluma cultivar is the most exploited by farmers (RAMOS, et al., 2010) due to the characteristics of its fruit such as color, pleasant smell and taste, mineral composition and richness in lycopene, attributes that guarantee the preference of various consumers (LIMA et al., 2010; OLIVEIRA et al., 2015).

Despite its socio-economic and food significance, studies are incipient on the management of saline water associated with potassium fertilization in guava growing conditions in the semi-arid Northeast of Brazil. This technique could be an important strategy for reducing the effects of the stress caused by high salt concentrations, since potassium plays a number of roles in the plant, particularly in controlling cell turgidity, activating enzymes involved in respiration and photosynthesis, regulating the processes of stomata opening and closing, carbohydrate transport, transpiration, resistance to frost, drought, salinity and disease; increasing resistance to lodging and being directly associated with the quality of agricultural products (DAVIS et al., 1997).

In addition, it can increase plant tolerance to salinity by inducing an increase in the K/Na, Ca/Na and NO/Cl ratios in plants (CUARTERO & MUNOZ, 1999), especially in the case of guava, which has potassium as one of the most demanded macronutrients (FRANCO et al., 2007). In this context, the aim of this study was to evaluate the formation of guava rootstock cv. Paluma as a function of the electrical conductivity of the irrigation water and doses of potassium.

2. MATERIAL AND METHODS

1.1 Characterization of the experimental area

The experiment was carried out in the greenhouse of the Center for Agrifood Sciences and Technology of the Federal University of Campina Grande, Pombal-PB Campus, from March 2015 to December 2015, located at the geographical coordinates of 6° 47'03" S, 37° 49'15" W and 193 m above sea level. The municipality's climate, according to the Koppen classification (BRASIL, 1972), is of the BSh type, i.e. hot semi-arid, with average annual rainfall of 750 mm and evaporation of 2000 mm.

1.2 Experimental design and treatments

The experimental design used was a randomized block design, with the treatments arranged in a 5 x 4 factorial scheme, referring to five levels of irrigation water salinity (ECa = 0.3; 1.1; 1.9; 2.7 and 3.5 dS m^{-1}) associated with four doses of potassium fertilization [(70, 100, 130 and 160% of K, corresponding to 508.2; 726; 943.8 and 1.161.6 mg of K dm^{-3} of substrate)], with each plot consisting of two plants, totaling 80 plots. The doses of K used were based on the absorption rate of this element in the formation phase of guava seedlings determined in hydroponics by Franco et al. (2007), with 726 mg of K dm^{-3} of substrate being the dose corresponding to 100% of the recommendation.

1.3 Description of treatments

The irrigation water was prepared by adding NaCl, CaCl$_2$ .2H$_2$ O and MgCl$_2$ .6H$_2$ O to water from the local supply system (Pombal-PB), which had an electrical conductivity of 0.3 dS m^{-1} , maintaining equivalent proportions of 7:2:1 for Na:Ca:Mg, the quantity of which (Q) was determined, based on Richards (1954), according to Eq. 1, which was calibrated using a portable conductivity meter and then stored in plastic containers with a capacity of 200 liters (Figure 1) to avoid evaporation and/or contamination by external agents.

Q (mg L^{-1}) = 640 x ECa Eq. (1).

In which:

Q = Amount of salt to be applied (mg L $)^{-1}$

ECa = represents the desired value of the water's electrical conductivity (dS m $)^{-1}$.

Figure 1. Plastic containers with a capacity of 200 L containing water of different salinities.

1.4 Rootstock production

The rootstocks were produced using polyethylene plastic bags with a capacity of 1.23 dm^{-3} as containers, perforated at the base for free water drainage. The bags were placed on metal benches, at a height of 0.8 m from the ground (Figure 2 A and B), to facilitate cultivation and application of the treatments.

Figure 2 - Overview of the experiment at the beginning (A) and at 170 days after emergence (DAE).

The substrate was made up of soil classified as Eutrophic Fluvial Neossolo with a sandy loam texture (EMBRAPA, 2013), fine sand and tanned cattle manure in the proportions of 82. 15 and 3% respectively, according to studies previously developed by Silva et al. (2015), whose physical and chemical characteristics (Table 1) were determined according to Claessen (1997).

Table 1. Physical and chemical characteristics of the substrate used in the experiment,

before the treatments were applied.

Textural classification	Apparent density g cm^{-3}	Total porosity %	Organic matter g kg^{-1}	P mg dm^{-3}	Ca^{+2} --	Mg^{2} + cmot	Na+ $_c$dm^{-3}	K+
							Assortative complex	
Sandy loam	1,38	47,00	32	17	5,4	4,1	2,	210,28

Saturation extract

pHesCEes dS m^{-1}	Ca^{+2} Mg^{+2}	K+	Na+ Cl --- mmol dm$_c^{-3}$	SO4^{2-}	CO3^{2-}	HCO3^{-} ---	Saturation %
7,411 ,21	2,503 ,75	4,74	3,027 ,50	3,10	0,00	5,63	27,00

pHes = pH of the substrate saturation extract;ECes = Electrical conductivity of the substrate saturation extract at 25°C

During the filling of the bags, 100 mg of P dm^3 of soil was incorporated into the substrate, using crushed simple superphosphate as a source, according to the recommendations of Corrêa et al. (2003) for the production of guava seedlings cv. Paluma.

The Paluma cultivar was used in this research because it is a genetic material adapted to the soil and climatic conditions of the Brazilian Northeast and is one of the most cultivated in Brazil (DIAS et al., 2012), due to its ease of access, high productivity, vigor, suitability for *fresh and* industrial consumption (PEREIRA et al., 2003), tolerance to pests and diseases, especially rust (*Puccinia psidii* Wint.) (MANICA et al., 2001).

The guava cv. Paluma seeds used to form the rootstocks for the experiment were obtained from a commercial orchard in the municipality of Aparecida/PB. Plants were standardized according to their vigor, freedom from pests and health (FACHINELLO et al., 2005). Fruits that were physiologically ripe, healthy and of a homogeneous size were collected, then pulped, washed in running water and left to dry in the shade under paper towels for 3 days. The seeds were sown equidistantly, at a rate of four seeds per bag, to a depth of 1.0 cm and, when the plants had an average of two pairs of true leaves, they were thinned out, leaving one plant per container, the most vigorous.

Cultivation consisted of manual weeding, superficial scarification of the substrate to remove compacted layers and pruning of lateral branches, since no pests and/or diseases were found.

1.5 Application of treatments

The soil was kept at near field capacity by means of a water balance in the substrate (BERNARDO et al., 2006), whose irrigation water had a low ECa level (0.3 dS m^{-1}) until the treatments were applied (40 days after seedling emergence - DAE). The irrigation events were carried out manually and took place in the early morning (8 a.m.) and late afternoon (5 p.m.). The volume of water applied was determined using the drainage lysimetry method, obtained by the difference between the volume applied and the volume drained from the previous irrigation, plus a leaching fraction of 0.15 (BERNARDO et al., 2006), in order to reduce the salinity of the substrate's saturation extract. At each irrigation event, the ECa was measured according to the pre-established treatments using a portable conductivity meter.

The bags had two holes in the base to allow drainage and plastic bottles were installed below them to monitor the volume of water drained and estimate the plant's water consumption.

Potassium fertilization began at 40 DAE and was divided into 24 equal weekly applications. The fertilizer used as a potassium source was potassium nitrate KNO$_3$ (14% Ne 48% K), applied manually using a Becker, simulating fertigation with water with an electrical conductivity of 0.3 dS m^{-1}, individually on each plot, according to the treatments.

In addition, 24 nitrogen fertilizations were carried out at weekly intervals, using urea

(45% N) as the source, according to the recommendations of Dias et al. (2012) for guava rootstocks propagated by herbaceous cuttings, whose dose was 773 mg of N dm^{-3} of substrate, considering the percentage of N (14%) provided by potassium nitrate.

1.6 Variables analyzed

The growth of guava rootstock cv. "Paluma" was assessed at 120 and 225 DAE, based on the stem diameter (mm), obtained at a height of 3 cm from the neck of the plant (Figure 3A); plant height (cm), by measuring the distance between the ground and the point of insertion of the apical meristem; the number of leaves, by counting the leaves with the leaf blade fully open; leaf area, determined by measuring the main vein of the leaves with the leaf blade fully open (Figure 3B), as recommended by Lima et al. (2012), considering Eq. 2.

$$AF = 0.3205 * C^{2,0412} \qquad \text{Eq. 2.}$$

In which:

AF= leaf area (cm)^2

C= length of the leaf's main vein (cm).

The absolute growth rate of plant height (ACRH) and its relative growth rate (ACRH) were determined during the period 60 - 225 DAE according to the methodology described by Benincasa (2003), as shown in Eqs. 3 and 4.

$$TCAap = (AP2-AP1) / (t2-t1) \qquad \text{Eq. 3.}$$

$$TCRap = (\ln AP2 - \ln AP1) / (t2-t1) \quad \text{Eq. 4.}$$

In which:

TCAap = absolute growth rate of plant height (cm day)$^{-1}$,

TCRap = Relative plant height growth rate (cm cm^{-1} day)$^{-1}$,

AP1 = plant height (cm) at time t1,

AP2 = plant height (cm) at time t2,

ln = natural logarithm.

The specific leaf area (SFA) was measured at 225 DAE, as described by Benincasa (2003), according to Eq. 5.

$$AFE = AF/FSFEq \qquad\qquad . 5.$$

In which:

AFE = specific leaf area, in $cm^2 g^{\wedge 1}$;

AF = leaf area, in cm^2 ;

FSF = leaf dry mass, in g.

Figure 3: Determination of stem diameter (A) and main vein length (B) of guava rootstocks cv. Paluma under salt stress.

2. 7Statistical analysis

The variables were evaluated by analysis of variance using the F test (1 and 5% probability) and, in cases of significant effect, linear and quadratic polynomial regression analysis was carried out using the SISVAR statistical software (FERREIRA, 2011). The regression was chosen according to the best fit based on the coefficient of determination (R^2) and taking into account a probable biological explanation for the treatments. Due to the heterogeneity of the data seen in the coefficient of variation values (Table 2 and 3), it was necessary to carry out an exploratory analysis of the data, transforming the data into $\sqrt{x}$ for the variables number of leaves, leaf area and specific leaf area.

3. RESULTS AND DISCUSSION

The results of the analysis of variance (Table 2) showed that there was a significant effect ($p \leq 0.05$) at 225 DAE of the levels of salinity in the irrigation water on all the growth variables studied. In relation to the doses of potassium, there was a significant difference ($p \leq 0.05$) for plant height (PH) at 120 and 225 DAE. There was no significant interaction between the factors salinity levels and potassium doses (S x DK) for any variable.

Table 2. Summary of the analysis of variance for plant height (PH), stem diameter (SD), number of leaves (NF) and leaf area (LA) of guava rootstocks cv. Paluma irrigated with salinized water and different doses of potassium fertilizer, at 120 and 225 DAE.

Source of variation	GL	Mean squares							
		AP		DC		NF		AF	
		120	225	120	225	120	225	120	225
Salt levels (S)	4	$14{,}06^{ns}$	$71{,}35*$	$0{,}11^{ns}$	$0{,}66*$	$4{,}64^{ns}$	$73{,}18*$	$4890{,}37^{ns}$	$104051{,}08*$
Linear Reg.	1	$4{,}52^{ns}$	$3{,}75^{ns}$	$0{,}30^{ns}$	$2{,}56^{**}$	$3{,}90^{ns}$	$172{,}22^{*}$	$956{,}87^{ns}$	$64347{,}67^{ns}$
Quadratic rule	1	$2{,}81^{ns}$	$197{,}81^{*}$	$0{,}00^{ns}$	$0{,}02^{ns}$	$0{,}11^{ns}$	$33{,}01^{ns}$	$2402{,}40^{ns}$	$342731{,}88^{**}$
Doses of K (DK)	3	$35{,}85^{*}$	$144{,}91^{*}$	$0{,}03^{ns}$	$0{,}20^{ns}$	$1{,}16^{ns}$	$27{,}18^{ns}$	$5247{,}57^{ns}$	$19329{,}62^{ns}$
Linear Regulation	1	$9{,}64^{ns}$	$385{,}14^{**}$	$0{,}01^{ns}$	$0{,}55^{ns}$	$0{,}81^{ns}$	$0{,}00^{ns}$	$682{,}41^{ns}$	$39432{,}03^{ns}$
Quadratic rule	1	$42{,}19^{*}$	$28{,}20^{ns}$	$0{,}01^{ns}$	$0{,}04^{ns}$	$1{,}25^{ns}$	$70{,}31^{ns}$	$3962{,}95^{ns}$	$15997{,}99^{ns}$
S*DK interaction	12	$23{,}28^{ns}$	$107{,}42^{ns}$	$0{,}11^{ns}$	$0{,}10^{ns}$	$5{,}01^{ns}$	$40{,}72^{ns}$	$10428{,}31^{ns}$	$21913{,}91^{ns}$
Blocks	3	$104{,}12^{**}$	$439{,}17^{**}$	$0{,}09^{ns}$	$0{,}27^{ns}$	$22{,}50^{**}$	$25{,}24^{ns}$	$42059{,}54^{**}$	$53475{,}45^{ns}$
CV (%)		7,87	9,20	7,28	9,11	7,16	17,72	13,36	28,00

ns,**, * respectively, not significant, significant at $p \leq 0.01$ and $p \leq 0.05$.

Increasing the salinity of the irrigation water negatively affected the height of the guava plants cv. Paluma at 225 DAE and according to the regression equation (Figure 4A) there was a quadratic response, the highest value of which (86.01 cm) was obtained when the guava plants were irrigated with ECa up to 1.8 dS m^{-1} , followed by a decrease in this variable, with the lowest estimated value of 81.95 cm reached when using water with a salinity of 3.5 dSm^{-1} . This reduction in plant height may be related to water deficiency, due to the osmotic effect, which can cause morphological and anatomical changes in plants, impairing water absorption and the rate of transpiration (SILVA et al., 2008), 2008). According to Santos et al. (2012), this reduction in height growth can be attributed to the fact that, in order to adjust osmotically, the plant releases a certain amount of energy to accumulate sugars, organic acids and ions in the vacuole.

Silva et al. (2015), evaluating the formation of guava rootstocks at 80 DAE, found a decline in height, with a reduction of 4.84% per unit increase in ECa. Similarly, Cavalcante et al. (2010) found an 11% decrease in the height of guava seedlings of the Paluma variety, per unit increase in ECa at 70 days after sowing, evaluating ECa levels from 0.5 to 4.0 dS m^{-1} .

The increase in the potassium dose led to a reduction in guava plant height in the evaluation carried out at 120 DAE and according to the regression equation (Figure 4B), the model that best fitted the data was the quadratic one, with the maximum value (45.87 cm) for plant height being obtained when using a dose of 110% K; a result that is close to that of Franco et al. (2007), who determined the macronutrient absorption rate for guava seedlings in hydroponics. However, during the 225 DAE period, guava plant height decreased linearly with the increase in potassium doses, with a reduction of 2.1% per 30% increase in K doses. When comparing the height of plants fertilized with 160% K, there was a reduction of 6.42% (5.88 cm) compared to those cultivated with the lowest dose of potassium (70% K). According to Satti & Lopez (1994), an increase in the dose of K does not always result in beneficial effects for the plants, and the salinity caused by high concentrations of K may even be more harmful than that caused by high concentrations of salts such as sodium and chloride, which may have occurred in this study due to the longer exposure time of the plants to saline stress.

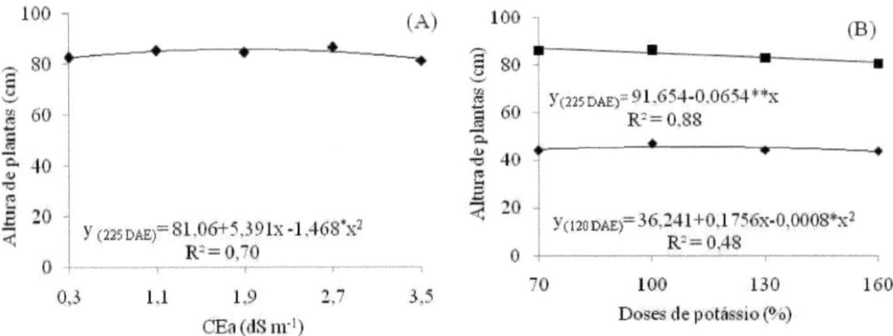

Figura 4. Plant height of guava rootstocks cv. Paluma, as a function of the electrical conductivity of the irrigation water - ECa at 225 DAE (A) and potassium doses at 120 and 225DAE (B).

Figure 5 shows a decreasing linear effect on guava stem diameter over the 225 DAE period as a function of the increase in irrigation water salinity, with a decrease of 2.83% per unit increase in ECa, in other words, the plants that were irrigated with an ECa of 3.5 dS m⁻¹ showed a decrease in CD of 9.05% (0.49 mm) compared to those that were irrigated with an ECa of 0.3 dS m⁻¹ ; This decline characterizes the negative effect of osmotic and hydric potential on the growth of guava plants during the rootstock formation phase. Souza et al. (2016), when studying the formation of 'Crioula' guava rootstock under irrigation with salinized water (ECa: 0.3 dS m⁻¹ to 3.5dS m⁻¹) and nitrogen fertilization, observed at 190 DAE that irrigation with water of ECa 3.5dS m⁻¹ caused a decrease of 13.28% compared to the lowest saline level (0.3 dS m⁻¹).According to Guimarâes et al, (2009), the diameter of the stem is an important characteristic, since its higher value provides vigor, robustness and resistance to the plant.

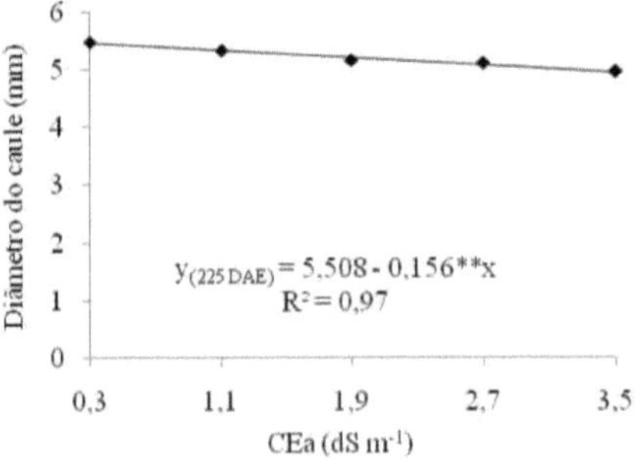

Figura 5. Stem diameter of guava rootstock cv. Paluma, as a function of the electrical conductivity

The number of leaves on the Paluma guava tree decreased linearly as the salinity of the irrigation water increased and, according to the regression study (Figure 6A), there was a 6.68% decrease per unit increase in the ECa at 225 DAE, resulting in a 21.38% decrease in the number of leaves on the plants irrigated with water with an ECa of 3.5 dSm^{-1} , compared to the plants that were irrigated with an ECa of 0.3 dS m^{-1} . In crop conditions under saline stress, it is common for there to be morphological and anatomical changes in the plants that promote a reduction in transpiration as an alternative to maintaining a lower need for water absorption; one of these adaptations is a reduction in the number of leaves (OLIVEIRA et al., 2013).

The leaf area of guava cv. Paluma was significantly affected by the increase in the electrical conductivity of the irrigation water at 225 DAE, fitting the quadratic regression model (Figure 6B), whose highest value for LA (456,56 cm^2) was reached at the ECa level of 1.7 dS m^{-1} , which then decreased as the salinity of the irrigation water increased, with the lowest value being obtained in the plants irrigated with the highest ECa level of 3.5 dS m^{-1} (254.35 cm^2). The PA of plants grown in saline environments is one of the most affected variables, the decrease in which is possibly due to the accumulation of salts in the leaves (MUNNS et al., 2006), and must be related to one of the plant's adaptation mechanisms to saline stress, reducing the transpiring surface while maintaining a high water potential in the plant by reducing transpiration (TESTER & DEVERPONTE, 2003).

In general, there was a negative effect of irrigation water salinity on the NF and AF of Paluma guava rootstocks. Silva et al. (2015), evaluating the growth and quality of guava rootstocks under irrigation with salinized water and nitrogen doses at 160 DAE, found a reduction in the number of leaves of 9.86% per unit increase in ECa, results similar to the present study, however, at 80 DAE the results were different in relation to leaf area, whose behavior was linearly decreasing with a reduction of 9.26% per unit increase in ECa.

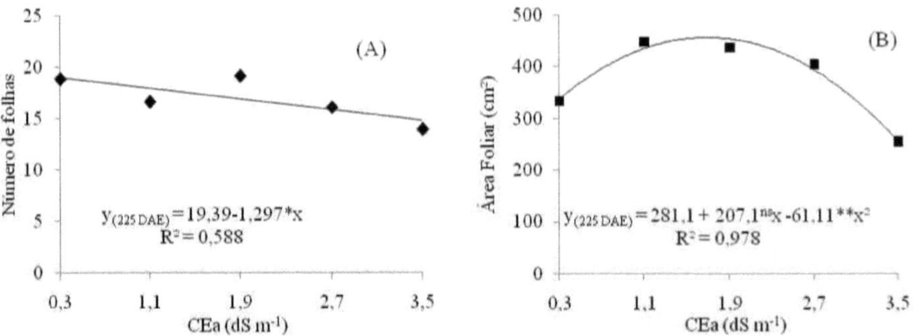

Figura 6. Number of leaves - NF (A) of guava rootstocks cv. Paluma and Leaf Area - AF (B), as a function of the electrical conductivity of the irrigation water - CEa, at 225 DAE.

According to the summary of the analysis of variance (Table 3), there was a significant effect ($p \leq 0.05$) of the salt levels in the irrigation water on the absolute growth rate for plant height (TCAap) between 60 and 225 DAE. However, the doses of potassium and the interaction between the factors (S x DK) did not significantly influence any of the variables studied.

Table 3. Summary of the analysis of variance for absolute growth rate for plant height (TCAap), relative growth rate for plant height (TCRap) during the period from 60 to 225 DAE and specific leaf area (AFE) at 225 DAE of guava rootstocks cv. Paluma irrigated with salinized water and different doses of potassium.

Source of variation	GL	Mean squares		
		TCAap	TCRap	AFE
		60 - 225	60 - 225	225
Salt levels (S)	4	$0{,}003^{*}$	$7{,}16^{ns}$	$23353{,}64^{ns}$
Linear Reg.	1	$0{,}0002^{ns}$	$0{,}00^{ns}$	$2083{,}18^{ns}$
Quadratic rule	1	$0{,}010^{*}$	$^{-6ns}$	$25417{,}07^{ns}$
Doses of K (DK)	3	$0{,}004^{ns}$	$4{,}59^{ns}$	$10151{,}04^{ns}$
Linear Reg.	1	$0{,}010^{ns}$	$_{10}{-6ns}$	$25766{,}83^{ns}$
Quadratic rule	1	$0{,}002^{ns}$	$0{,}00^{ns}$	$4327{,}53^{ns}$
S*DK interaction	12	$0{,}002^{ns}$	$3{,}63^{ns}$	$7620{,}16^{ns}$
Blocks	3	$0{,}010^{**}$	$6{,}04^{ns}$	$9327{,}74^{ns}$

CV (%)	12,53	8,80	22,39

ns, **, * respectively not significant, significant at p ≤0.01 and p ≤ 0.05

Irrigation water salinity negatively affected the absolute growth rate for plant height (Table 3) and, according to the regression equation (Figure 7), the model to which the data fitted best was the quadratic one, with the maximum TCAap value (0.37009 cm day^{-1}) obtained at the ECa level of 1.9 dS m^{-1} , from this level there was a decrease in this variable, reaching the lowest value when irrigated with the highest saline level (3.5 dSm^{-1}). The growth inhibition observed through the TCAap may have been caused by excess salts in the substrate, which according to NOBRE et al. (2014), glycophyte plants such as guava, when exposed to salt concentrations above the threshold level, reduce water absorption, the rate of photosynthesis and, consequently, their growth.

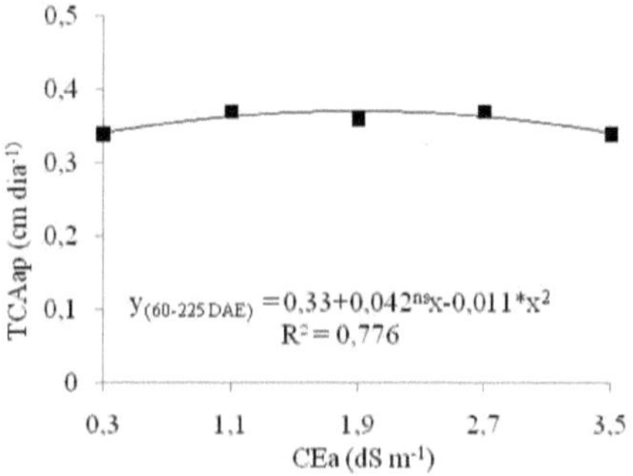

Figura 7. Absolute growth rate for plant height (TCAap) of guava cv. Paluma during the period from 60 to 225 DAE as a function of the electrical conductivity of the water - ECa.

Although sodium and chloride salts cause toxicity to guava rootstocks, affecting their physiological quality and compromising the absorption of water and nutrients, due to the reduction in the osmotic potential of the substrate and the competition between specific ions, causing disturbances in plant growth (SOUSA et al., 2011; SA et al., 2013) and potassium is one of the nutrients most required by guava trees (FRANCO et al., 2007), 2007), in this study, the salinity of the irrigation water, doses of potassium and the interaction between the factors did not have a significant effect (p <0.05), according to the F test, on the relative growth rate for plant height (TCRap) during the period from 60 to 225 DAE and the specific leaf area (AFE) at 225 DAE of guava rootstocks cv. Paluma (Table 3). Paluma (Table 3), which showed average values of 0.0074 cm cm^{-1} day^{-1} and 162.12 cm^2 g^{-1} , respectively.

4. CONCLUSIONS

1.   Irrigation with ECa water of up to 1.9 dS m^{-1} made it possible to form guava rootstock cv. Paluma with an acceptable reduction in its growth;

2.   The potassium dose of 798.6 mg of K dm^{-3} promoted the greatest growth in height of the guava rootstock cv. Paluma at 120 days after emergence;

3.   Increasing doses of K did not mitigate the harmful effects of salts on the growth variables evaluated;

4.   The effects of irrigation water salinity and potassium fertilization on the growth of guava rootstocks are independent.

5. BIBLIOGRAPHICAL REFERENCES

ALMEIDA, O. A. **Quality of irrigation water.** Cruz das Almas: Embrapa Cassava and Fruit Growing, 2010. 227p.

BENINCASA, M. M. P. **Plant growth analysis, basics.** 2 ed. Jaboticabal: FUNEP, 2003. 41p.

BERNARDO, S.; SOARES, A. A.; MANTOVANI, E. C. **Manual de irrigaçâo.** 8 ed., Viçosa: UFV, 2006. 625 p.

BEZERRA, J. D.; PEREIRA, W. E.; SILVA J. da M.; RAPOSO, R. W. C. Growth of two genotypes of yellow passion fruit under salinity conditions. **Revista Ceres,** Viçosa, v. 63, n.4, p. 502-508, 2016.

BRAZIL. Ministry of Agriculture. Research and Experimentation Office. Pedology and Soil Fertility Team. **I Exploratory and reconnaissance survey of soils in the state of Paraiba. II Interpretaçâo para uso Agricola dos Solos do Estado da Paraiba.** Rio de Janeiro: 1972. 638p. (Technical Bulletin, 15; SUDENE, Pedological Series, 8).

CAVALCANTE, L. F.; CAVALCANTE, I. H. L.; PEREIRA, K S. N.; DE OLIVEIRA, F. A.; GONDIM, S C.; DE ARAÙJO, F A. R. Germination and initial growth of guava plants irrigated with saline water. **Revista Brasileira de Engenharia Agricola e Ambiental,** Campina Grande, v. 9, n. 4, p. 515-519, 2005.

CAVALCANTE, L. F; VIEIRA, M. S.; SANTOS, A. F.; OLIVEIRA, W. M.; NASCIMENTO, J. A. M. Saline water and liquid bovine manure in the formation of guava seedlings cultivar Paluma. **Revista Brasileira de Fruticultura,** Jaboticabal, v. 32, n. 01, p. 251-261, 2010.

CLAESSEN, M. E. C. (org.). **Manual of soil analysis methods.** 2. ed. rev. atual. Rio de Janeiro: Embrapa-CNPS, 1997. 212p. (Embrapa-CNPS. Documentos, 1).

CORRÊA, M. C. M.; PRADO, R. M.; NATALE, W.; PEREIRA, L.; BARBOSA, J. C. Responses of guava seedlings to doses and modes of application of phosphate fertilizer. **Revista Brasileira de Fruticultura,** Jaboticabal, v. 25, n. 1, p. 164-169,

2003.

DAVIS, R. M.; SUBBARAO K. V.; RAID R. N.; KURTZ E. A. 1997. *Compendium of lettucediseases.*California: Academic Press. 79p.

DIAS, M. J. T.; SOUZA, H. A.; NATALE, W.; MODESTO, V. C.; ROZANE, D. E. Nitrogen and potassium fertilization for guava seedlings in a commercial nursery. **Semina: Ciências Agràrias**, Londrina, v. 33, supplement 1, p. 2837-2848, 2012.

CUARTERO, J.; MUNOZ, R. F. Tomato and salinity. **Scientia Horticulturae**, Amsterdam, v.78, n.1-4, p.83-125, 1999.

BRAZILIAN AGRICULTURAL RESEARCH Corporation - EMBRAPA. National Soil Research Center. **Brazilian Soil Classification System**. 3. ed.Brasilia, 2013. 353p.

FACHINELLO, J. C.; NACHTIGAL, J.C.; HOFFMANN, A. Propagation by seeds. iN: FACHiNELLO, J. C.; HOFFMANN, A.; NACHTiGAL, J.C. (Ed.). **Propagation of fruit plants.** Brasilia, DF: Embrapa Informaçao Tecnològica. 2005. P. 57-67.

FERREIRA, D. F. Sisvar: a computer system for statistical analysis. **Ciência e agrotecnologia**, v. 35, n. 6, p. 1039-1042, 2011.

FRANCO, F. C.; PRADO, R. M.; BRACHIROLLI, L. F.; ROZANE, D. E. Growth curve and macronutrient uptake in guava seedlings. **Revista Brasileira de Ciência do Solo**, Viçosa, v. 31, n. 6, p. 1429-1437, 2007.

GUIMARAES, A. S.; MACEDO, B. N. E.; COSTA, S. G. Increasing sources and doses of organic and mineral fertilizers in the initial growth of pinhao manso. **Mens agitat**, Boa Vista, v. 04, n. 1, p. 17-22, 2009.

HOLANDA, J. S.; AMORIM, J. R. A.; FERREIRA NETO, M.; HOLANDA, A. C. Quality of water for irrigation. In: GHEYI, H. R.; DIAS, N. S.; LACERDA, C. F de (Ed). **Salinity management in agriculture:** basic and applied studies. Fortaleza: INCTSal, 2010. p. 43-61.

LIMA, M. S.; PIRES, E. M. F.; MACIEL, M. I. S.; OLIVEIRA, V. A. Quality of minimally processed guava with different types of cut, sanification and packing.

Ciência e Tecnologia de Alimentos, v.30, n.1, p.79-87, 2010.

LIMA, L. G. S.; ANDRADE, A. C.; SILVA, R. T. L.; FRONZA, D.; NISHIJIMA, T. **Mathematical models for estimating the leaf area of guava trees (*Psidium guajava* L.).** In: 64ª SBPC ANNUAL MEETING. Sao Luiz: UFMA, 2012.

LIMA, G. S.; NOBRE, R. G.; GHEYI, H. R. SOARES, L. A. A.; SILVA, A. O. Production of papaya cultivated with saline water and nitrogen doses. **Revista Ciência Agronômica,** Fortaleza, v.46, n.1, p.1-10, 2015.

MANICA, I.; ICUMA, I. M.; JUNQUEIRA, N. T. V.; SALVADOR, J. O.; MOREIRA, A.; MALAVOLTA, E. **Goiaba: From planting to consumer: Production technology, post-harvest, marketing.** Porto Alegre: Cinco Continentes, 2001. 124 p.

MEDEIROS, J. F. de; NASCIMENTO, L. B. do; GHEYI, H. R. Soil-water-plant management in areas affected by salts. In: GHEYI, H. R.; DIAS, N. S.; LACERDA, C. F de (Ed). **Salinity management in agriculture:** basic and applied studies. Fortaleza: INCTSal, 2010. p. 279-302.

MEDEIROS, J. F. DE; LISBOA, R. A.; OLIVEIRA, M.; SILVA JÚNIOR, M.J.; ALVES, L. P. Characterization of groundwater used for irrigation in the Chapada do Apodi melon-producing area. **Revista Brasileira Engenharia Agricola e Ambiental,** Campina Grande, v. 7, n.3, p. 469- 472, 2003.

MUNNS, R.; JAMES, R.A.; LĂUCHLI, A. Approaches to increasing the salt tolerance of wheat and other cereals. **Journal of Experimental Botany**, Oxford, v.57, p.1025-1043, 2006.

NEVES, A. L. R.; LACERDA, C. F.; GUIMARÂES, F. V. A.; Hernandez, F. F. F.; Silva, F. B.; Prisco, J. T.; Gheyi, H. R. Biomass accumulation and nutrient extraction by string bean plants irrigated with saline water at different stages of development. **Revista Ciência Rural,** Santa Maria, v. 39, n. 3, p. 758-765, 2009.

NOBRE, R. G.; LIMA, G. S.; GHEY, H. R.; SOARES, L. A. A.; SILVA, A. O.

Growth, consumption and efficiency of water use by papaya under saline and nitrogen stress. **Revista Caatinga.** Mossoró, v.27, n.2, p.148-158, 2014.

NUNES, J. C.; CAVALCANTE, L. F.; LIMA NETO, A. J.; SILVA, J. A.; SOUTO, A. G. L.; ROCHA, L. F. Humitec® and soil mulch on the initial growth of guava cv. "Paluma" in the field. **Revista Agro@mbiente On-line**, Roraima, v. 8, n. 1, p. 89-96, 2014.

OLIVEIRA, F. T. DE; HAFLE, O. M.; MENDONÇA, V.; MOREIRA, J. N.; PEREIRA JÙNIOR, E. B.; ROLIM, H. O. Responses of guava rootstocks under different sources and proportions of organic materials. **Comunicata Scientiae**, v.6, p.17-25, 2015.

OLIVEIRA DE, F. A.; MEDEIROS DE, J. F.; OLIVEIRA DE, M. K. T.; SOUZA, A. A. T.; FERREIRA, J. A.; SOUZA, M. S. Interaction between salinity and biostimulant in cowpea culture. **Brazilian Journal of Agricultural and Environmental Engineering.** Campina grande, v.17, n.5, p.465-471, 2013.

PEREIRA, F. M.; CARVALHO, C.; NACHTIGAL, J.C. Século XXI: a new dual-purpose guava cultivar. **Revista Brasileira de Fruticultura**, Jaboticabal, v. 25, n.3, p.498-500, 2003.

RAMOS, D. P.; SILVA, A. C.; LEONEL, S.; COSTA, S. M.; DAMATTO JÙNIOR, E. R. Fruit production and quality of the 'Paluma' guava tree, subjected to different pruning times in a subtropical climate. **Revista Ceres**, Viçosa, v. 57, n.5, p. 659-664, 2010.

RICHARDS, L. A. (ed.) **Diagnosis and improvement of saline and alkali soils**. Washington, United States Salinity Laboratory: 1954. 160p. (USDA. Agriculture Hand book, 60).

SANTOS, B. dos.; FERREIRA, P. A.; OLIVEIRA, F. G. de.; BATISTA, R. O.; COSTA, A. C.; CANO, M. A. O. Yield and physiological parameters of peanut as a function of salt stress. **Idésia**, v.30, n.2, p. 69-74, 2012.

SATTI, S. M. E.; LOPEZ, M. **Effect of increasing potassium levels for alleviating**

sodium chloride stress on the growth and yield of tomato. Communications Soil Science and Plant Analysis, v.25, n.15-16, p.2807-2823, 1994.

SILVA, E. M.; NOBRE, R. G.; GHEYEI, H. R.; PINHEIROS, F. W. A.; SOUZA, L. de P.; DIAS, A. S. **Growth and quality of guava rootstock under irrigation with salinized water and nitrogen doses.** In: III Inovagri International Meeting, Fortaleza, CE. 2015.

SILVA, S. M. S.; ALVES, A. N.; GHEYI, H. R. Development and production of two cultivars of papaya under saline stress, **Revista Brasileira de Engenharia Agricola e Ambiental,** Campina Grande, v.12, n.4, p.335-342, 2008.

SA, F. V. S; BRITO, M. E. B; MELO, A. S; ANTÔNIO NETO, P .; FERNANDES, P. D; FERREIRA, I. B. Production of papaya seedlings irrigated with saline water. **Revista Brasileira de Engenharia Agricola Ambiental**, Campina Grande, v.17, p.1047-1054, 2013.

SOARES, T. M.; SILVA, E. F. F.; DUARTE, S. N.; MELO, R. F.; JORGE, C. A.; BONFIM, S., E. M. Lettuce production using saline water in a hydroponic system. **Irriga**, Botucatu v.12, p.235-248, 2007.

SOUSA, A. B. O.; BEZERRA, M. A.; FARIAS, F. C. Germination and initial development of common cashew clones under irrigation with saline water. **Revista Brasileira de Engenharia Agricola e Ambiental**, Campina Grande, v.15, p.390-394, 2011.

SOUZA, L. P., NOBRE R. G., SILVA E. M.; LIMA, G. S.; PINHEIRO, F. W. A.; ALMEIDA L. L. de S. Formation of 'Crioula' guava rootstock under saline water irrigation and nitrogen doses. **Revista Brasileira de Engenharia Agricola e Ambiental,** Campina Grande, v.20, n.8, p.739-745, 2016.

TESTER, M.; DAVENPORT, R. Na^+ tolerance and Na^+ transport in higher plants. **Annals of Botany**, v. 91, n.3, p. 503-527, 2003.

CHAPTER II PHYTOMASS AND QUALITY OF GUAVA SEEDLINGS IRRIGATED WITH SALINE WATER AND POTASSIUM FERTILIZER

PHYTOMASS AND QUALITY OF GUAVA ROOTSTOCK IRRIGATED WITH SALINE WATER AND POTASSIUM FERTILIZATION

Abstract: In the semi-arid region of northeastern Brazil, the guava crop stands out due to its socio-economic and food importance, however, in this region the quality of irrigation water has high levels of salts which limits its rational exploitation. With this in mind, the aim of this study was to evaluate the formation of dry phytomass and the quality of guava rootstocks cv. Paluma irrigated with water of different salinities associated with doses of potassium in an experiment conducted using Eutrophic Fluvial Neosol with a sandy loam texture under greenhouse conditions in the municipality of Pombal - PB. The experimental design used was randomized blocks, in a 5 x 4 factorial scheme, whose treatments resulted from the combination of five levels of electrical conductivity of the irrigation water (ECa = 0.3; 1.1; 1.9; 2.7 and 3.5 dS m^{-1}) and four doses of potassium (70, 100, 130 and 160% of K), with the 100% K dose corresponding to 726 mg of K dm^{-3} of substrate, with four replications and two useful plants per plot. Irrigation with water of ECa 1.9 dS m^{-1} promotes an acceptable 10% reduction in phytomass production and quality of Paluma guava rootstocks; the dose of 508.2 mg of K dm^{-3} of substrate favors the accumulation of dry phytomass of Paluma guava stems at 225 DAE; there was no significant interaction (salt x doses of K) on the variables studied.

Keywords: *Psidium guajava* L., salt stress, potassium.

1 INTRODUCTION

The guava tree (*Psidium guajava* L.) is among the most expressive fruit species in Brazilian agribusiness, with great potential for expansion, especially in the northeast region, due to the favorable soil and climate conditions (IBGE, 2016), with the Paluma cultivar standing out as the most widespread in Brazil and preferred by the most diversified consumer markets (RAMOS et al., 2010).

Despite the good adaptation of this fruit tree to northeastern Brazil, in this region there are limitations in water resources involving both quantitative and qualitative aspects, especially with regard to the presence of salts in irrigation water (MEDEIROS et al., 2003). However, due to the growing demand for food, it has become necessary to use lower quality water, such as saline water. However, the use of saline water for irrigation raises the levels of salts in the soil solution, causing negative effects on plants by inhibiting germination, emergence, growth and biomass accumulation (CAVALCANTE et al., 2010).

It should be noted that the effect of water salinity on crops varies between species, genotypes, salinity levels, fertilization and irrigation management, and soil and climate conditions (BRITO et al., 2014), and there is a need to identify tolerant genetic materials. In this sense, research has been carried out using salinized water in the northeast region, especially in the formation of guava seedlings (CAVALCANTE et al., 2010; SOUZA, et al., 2016), however, these studies are quite incipient when it comes to the interaction between saline levels and potassium fertilization, highlighting the importance of further studies on this fruit tree at this stage of development.

Therefore, it is necessary to adopt water and soil management strategies, such as mineral fertilization, in order to reduce the deleterious effects of high salt concentrations on plants (SA et al., 2015). In addition, fertigation has been a widely used technique in the northeastern region of Brazil (MEDEIROS et al., 2012), with potassium being one of the most used macronutrients due to its deficiency in most soils (MALAVOLTA, 2005), one of the nutrients most required by guava seedlings

(FRANCO et al., 2007), as well as having an osmoregulatory function (EPSTEIN & BLOOM, 2006).

Considering the scarcity of scientific results involving the tolerance of guava trees irrigated with saline water associated with increasing doses of potassium, this study aims to evaluate the interaction of these factors on the formation of phytomass and the quality of guava rootstocks cv. Paluma.

2. MATERIAL AND METHODS

1.1 Characterization of the experimental area

The experiment was carried out in the greenhouse of the Center for Agrifood Sciences and Technology of the Federal University of Campina Grande, Pombal-PB Campus, from March 2015 to December 2015, located at the geographical coordinates of 6°47'03" S, 37°49'15" W and an altitude of 193 m. According to the Koppen classification (BRASIL, 1972), the municipality's climate is of the BSh type, i.e. hot semi-arid, with average annual rainfall of 750 mm and evaporation of 2000 mm.

1.2 Experimental design and treatments

The treatments were arranged in randomized blocks, in a 5 x 4 factorial scheme, referring to five levels of irrigation water salinity (ECa = 0.3; 1.1; 1.9; 2.7 and 3.5 dS m^{-1}) associated with four doses of potassium fertilization [(70, 100, 130 and 160% K, corresponding to 508.2; 726; 943.8 and 1. 161.6 mg K dm of substrate)], with each plot consisting of two plants.161.6 mg of K dm^{-3} of substrate)], with each plot consisting of two plants, making a total of 80 plots. The doses of K used were based on the absorption rate of this element in the formation phase of guava seedlings determined in hydroponics by Franco et al. (2007), with 726 mg of K dm^{-3} of substrate being the dose corresponding to 100% of the recommendation.

1.3 Description of treatments

The irrigation water was prepared by adding NaCl, $CaCl_2$.$2H_2$ O and $MgCl_2$.$6H_2$ O in water from the local supply system (Pombal-PB), which had an electrical conductivity of 0.3 dS m^{-1} , maintaining equivalent proportions of 7:2:1 for Na:Ca:Mg, whose quantity (Q) was determined, based on Richards (1954), according to Eq. 1, which were calibrated using a portable conductometer and then stored in plastic containers with a capacity of 200 liters to prevent evaporation and/or contamination by external agents.

Q (mg L^{-1}) = 640 x ECa Eq. (1).

In which:

Q = quantity of salts to be applied (mg L)$^{-1}$

ECa = represents the desired value of the water's electrical conductivity (dS m).$^{-1}$

1.4 Rootstock production

To form the rootstocks, polyethylene plastic bags with a capacity of 1.23 dm^3 were used as containers, perforated at the base for free water drainage. The substrate was made up of soil classified as Eutrophic Fluvial Neosol with a sandy loam texture (EMBRAPA, 2013), fine sand and tanned cattle manure in the proportions of 82, 15 and 3% respectively, according to previous work by Silva et al. (2015), whose physical and chemical characteristics (Table 1) were determined according to Claessen (1997).

Table 1. Physical and chemical characteristics of the substrate used in the experiment before the treatments were applied.

Textural classification	Apparent density $_{-3}$ g cm %	Total porosity	Organic matter g kg^{-1}	P mg dm^{-3}	Assortative complex			
					Ca $+^2$ ---	Mg $+^2$ Na+ ---- cmol dm $_c^{-3}$ ---		K+ ---
Sandy loam	1,38	47,00	32	17	5,4	4,	12,21	0,28

Saturation extract										
pHes	CEes dS m^{-1}	Ca $+^2$	Mg $+^2$	K+	Na+ mmolcdm^{-3} -	Cl$^-$	SO?	CO3^{2-}	HCO3- ---	Saturation %
7,41	1,21	2,50	3,75	4,743	,027	,50	3,10	0,00	5,63	27,00

pHes = pH of the substrate saturation extract; ECes = Electrical conductivity of the substrate saturation extract at 25°C

While the bags were being filled, 100 mg of P dm^{-3} of soil was added to the substrate, using crushed simple superphosphate as a source, in accordance with the recommendations of Corrêa et al. (2003) for substrate in the production of guava cv. Paluma seedlings.

The Paluma cultivar was used in this research because it is a genetic material adapted to the soil and climatic conditions of the Brazilian Northeast and is one of the most cultivated in Brazil (DIAS et al., 2012), due to its ease of access, high productivity, vigor, suitability for *fresh* and industrial consumption (PEREIRA et al., 2003), tolerance to pests and diseases, especially rust (*Puccinia psidii* Wint.) (MANICA et al., 2001).

The guava seeds used to form the rootstocks for the experiment were obtained from a commercial orchard in the municipality of Aparecida/PB. Plants were standardized according to their vigour, freedom from pests and health (FACHINELLO et al., 2005). Fruits that were physiologically ripe, healthy and of a homogeneous size were collected, then pulped, washed in running water and left to dry in the shade under paper towels for 3 days. The seeds were sown equidistantly, at a rate of four seeds per bag, to a depth of 1.0 cm and, when the plants had an average of two pairs of true leaves, one plant per container, the most vigorous, was thinned out.

Cultivation consisted of manual weeding, superficial scarification of the substrate to remove compacted layers and pruning of lateral branches, since no pests and/or diseases were found.

1.5 Application of treatments

The soil was kept at close to field capacity by means of a water balance in the substrate (BERNARDO et al., 2006), irrigated with water with a low ECa level (0.3 dS $_{m-1}$) until the treatments were applied (40 days after seedling emergence - DAE). Irrigation events were carried out manually and took place in the early morning (8am) and late afternoon (5pm). The volume of water applied was determined using the drainage lysimetry method, obtained by the difference between the volume applied and the volume drained from the previous irrigation, plus a leaching fraction of 0.15 (BERNARDO et al., 2006), in order to reduce the salinity of the substrate's saturation extract. At each irrigation event, the ECa was measured according to the pre-established treatments using a portable conductivity meter.

The bags had two holes in the base to allow drainage and plastic bottles were installed below them to monitor the volume of water drained and estimate the plant's water consumption.

Potassium fertilization began at 40 DAE and was divided into 24 equal weekly applications. The fertilizer used as a potassium source was potassium nitrate KNO$_3$ (14% Ne 48% K), with applications made manually using a Becker, simulating fertigation with water with an electrical conductivity of 0.3 dS m^{-1} , individually on each plot, according to the treatments.

In addition, 24 nitrogen fertilizations were carried out at weekly intervals, using urea (45% N) as the source, in accordance with the recommendations of Dias et al. (2012) for guava rootstocks propagated by herbaceous cuttings, whose dose was 773 mg of N dm^{-3} of substrate, taking into account the percentage of N (14%) provided by potassium nitrate.

1.6 Variables analyzed

To analyze the effect of the treatments on the phytomass production of guava rootstocks at 225 DAE, the plants were collected, the roots were washed to remove the adhered soil and each plant was divided into leaf, stem and root. This material was then placed in a previously identified paper bag and taken to the laboratory to determine the leaf area according to Lima et al. (2012), according to Eq. 2, then they were placed in an air circulation oven at 65 °C for 72 hours (Figure 1A), after which they were weighed on an analytical balance (Figure 1B) to determine the dry stem phytomass (DSF), leaf dry phytomass (FSF), root dry phytomass (FSR) and aerial part dry phytomass (FSPA) (stem + leaves). The sum of these phytomasses resulted in the total dry phytomass (FST).

$$AF = 0.3205 * C\ 1^{2,042} \qquad \text{Eq. (2).}$$

In which:

AF = leaf area (cm)^2

C = length of the leaf's main vein (cm).

Figure 1 - Material dried in an oven at 65°C (A) and weighing of the dry phytomass of guava cv. Paluma rootstocks (B).

We also studied the root to aerial part ratio (R/PA) (g g^{-1}) by dividing the dry phytomass of the root by the dry phytomass of the aerial part and the leaf area ratio (RAF) (cm^2 g^{-1}) (BENINCASA, 2003), using the ratio between the leaf area and the total dry phytomass of the plant, described in eq. 3.

RAF = AF/FST Eq. (3)

In which:

RAF = leaf area ratio (cm^2 g)$^{-1}$

AF= leaf area (cm)^2

FST = total plant dry phytomass (g)

The quality of the rootstocks was measured using the Dickson quality index (DQI) for seedlings, using the formula of Dickson et al. (1960), described by equation 4.

$$IQD = \frac{(FST)}{(AP/DC) + (FSPA/FSR)}$$

In which:

IQD = Dickson's quality index

FST = total plant dry phytomass (g)

AP = plant height (cm), DC = stem diameter (cm).

FSPA = dry phytomass of the aerial part of the plant (g)

FSR = plant root dry phytomass (g)

2.7 Statistical analysis

The variables were evaluated by analysis of variance using the F test (1 and 5% probability) and, in cases of significant effect, linear and quadratic polynomial regression analysis was carried out using the SISVAR statistical software (FERREIRA, 2011). The regression was chosen according to the best fit based on the coefficient of determination (R^2) and taking into account a probable biological explanation for the treatments. Due to the heterogeneity of the data seen in the coefficient of variation values (Tables 2 and 3), it was necessary to carry out exploratory data analysis, transforming the data into $\sqrt{x}$ for the variables leaf dry phytomass, root dry phytomass, shoot dry phytomass, total dry phytomass, root to shoot ratio, leaf area ratio and the Dicson quality index.

3. RESULTS AND DISCUSSION

The summary of the analysis of variance (Table 2) shows that there was a significant effect (p $\leq$ 0.01) of the salinity levels of the irrigation water on the dry phytomass of the stem (FSC), leaves (FSF), aerial part (FSPA) and root (FSR) at 225DAE. In relation to potassium doses, there were significant effects (p $\leq$ 0.05) only for stem dry phytomass. There was no significant interaction between the factors salinity levels and potassium doses (S x DK) for any of the variables studied.

Table 2. Summary of the analysis of variance for the dry mass of the stem (FSC), leaves (FSF), aerial part (FSPA) and root (FSR) of guava rootstock cv. Paluma at 225 DAE irrigated with water of different saline levels and fertilized with potassium.

Variation source	GL	Mean Square			
		FSC	FSF	FSPA	FSR
Salt levels (S)	4	$6,78^{**}$	$7,26^{**}$	$26,41^{**}$	$1,66^{**}$
Linear Reg.	1	$23,39^{**}$	$16,76^{**}$	$80,27^{**}$	$6,24^{**}$
Quadratic rule	1	$1,74^{ns}$	$9,19^{*}$	$19,17^{*}$	$0,03^{ns}$
Doses of K (DK)	3	$1,45^{*}$	$0,84^{ns}$	$4,27^{ns}$	$0,39^{ns}$
Linear Regulation	1	$3,60^{*}$	$1,44^{ns}$	$9,77^{ns}$	$0,16^{ns}$
Quadratic rule	1	$0,48^{ns}$	$0,00^{ns}$	$0,60^{ns}$	$0,00^{ns}$
Interaction (S*DK)	12	$0,67^{ns}$	$0,83^{ns}$	$1,82^{ns}$	$0,15^{ns}$
Blocks	3	$1,77^{*}$	$4,99^{*}$	$9,35^{*}$	$0,57^{*}$
CV (%)		20,16	28,96	14,53	16,15

ns, **, * respectively not significant, significant at p $\leq$ 0.01 and p $\leq$ 0.05

Increasing levels of irrigation water salinity negatively affected the dry phytomass of the stem and, according to the regression equation (Figure 2A), the model to which the data best fitted was linear, with decreases in SCF of 9,17% per unit increase in ECa, with a reduction in SCF of 29.35% (1.50 g per plant) being seen at 225 DAE in plants irrigated with water of 3.5 dS m^{-1} compared to plants irrigated with water of 0.3 dS m^{-1} . The reduction in guava FSC seen with increasing ECa is possibly associated with the osmotic effect of salts, toxic and nutritional effects resulting from the accumulation of salts in the plant's root zone, corroborating the results obtained by Souza et al. (2016), when evaluating the formation of 'Crioula' guava rootstock under irrigation with salinized water (ECa ranging from 0.3 to 3.5 dS m^{-1}) and nitrogen fertilization (DN = 541, 773, 1,005 and 1237 mg of N dm^{-3} of substrate) obtained a reduction of 7.6% per unit increase in the electrical conductivity of the irrigation water.

For the accumulation of dry phytomass in the leaves, there was a quadratic behavior as a function of the increase in salinity, so that, when evaluating the average values, the plants irrigated with water of ECa1.3 dS m^{-1} , obtained 2.96 g plant^{-1} (Figure 2B).In general, the guava rootstocks cv. Paluma suffered deleterious effects when irrigated with salinized water during the study period (225 DAE) in relation to the accumulation of phytomass in the stem and leaves, which was reflected in the loss of dry phytomass in the aerial part (Figure 3A). According to Silva et al. (2008), the reduction in biomass accumulation is a consequence of osmotic adjustment mechanisms under conditions of salt stress, including changes in the ion balance, water potential, stomatal closure and photosynthetic efficiency.

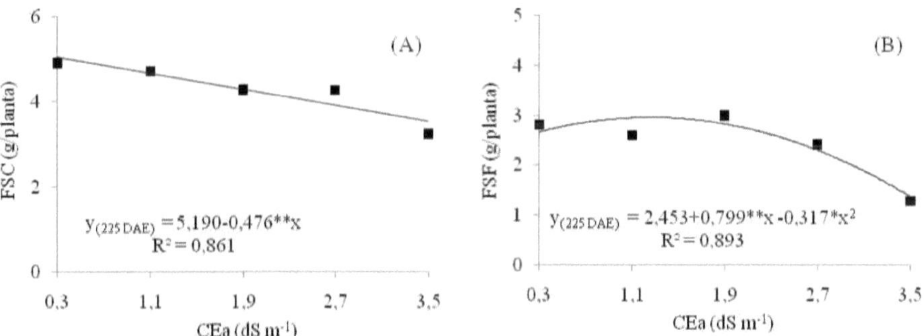

Figura 2. Dry phytomass of stems - FSC (A) and dry phytomass of leaves - FSF (B) of guava rootstocks cv. Paluma as a function of the electrical conductivity of the irrigation water - ECa at 225 DAE.

According to the regression equation (Figure 3A), the FSPA suffered a quadratic effect due to the increase in salinity of the irrigation water, with increasing values up to the ECa level of 0.9 dS m^{-1} , whose phytomass of the aerial part was 7.72 g, followed by a decrease in this variable, with the lowest estimated value of 4.72 g when using water with a salinity of 3.5 dS m^{-1} . This reduction in FSPA can be attributed to the lower number of leaves, leaf area and smaller stem diameter of the plants, and these effects of salinity were recorded by Souza et al. (2016).

The average values of root dry phytomass were higher in the treatments in which the plants were irrigated with water with lower saline levels. The reduction in root dry mass per unit increase in ECA was 12.51%, which represents a decrease of 40.04% (0.76 g per plant) in plants irrigated with ECA of 3.5 dS m^{-1} compared to plants irrigated with water of 0.3 dS m^{-1} (Figure 3B). The reduction in the dry phytomass of guava roots with increasing water salinity was similar to that observed by Gurgel et al. (2007), with the reductions being attributed to the osmotic and ionic effects caused by salinity.

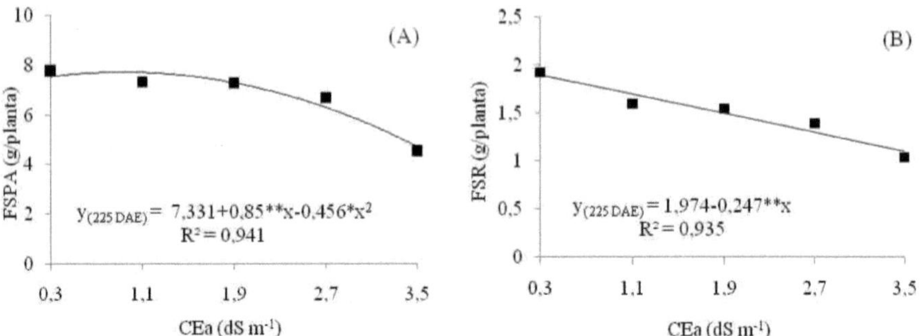

Figura 3. Dry phytomass of the aerial part - FSPA (A) and root - FSR (B) of guava rootstocks cv. Paluma as a function of the electrical conductivity of the irrigation water - ECa at 225 DAE.

The increase in K doses negatively affected SCF at 225 DAE, with the best fit of the data occurring in a linear regression equation (Figure 4), causing reductions of 3.59% for each 30% increase in the K dose, resulting in losses of 10.77% (0.5 g per plant) in plants fertilized with the 160% dose compared to those receiving the 70% K dose (Figure 4). According to Alcarde et al. (2007), the nutritional requirements of plants vary according to the stage of development, so when a higher amount is applied than that required by the species, antagonistic effects can occur, similar to those obtained by Souza et al. (2016) when assessing the formation of 'Crioula' guava rootstock irrigated with water of different salinity levels and nitrogen doses, they observed that increasing doses of nitrogen inhibited the formation of plant phytomass.

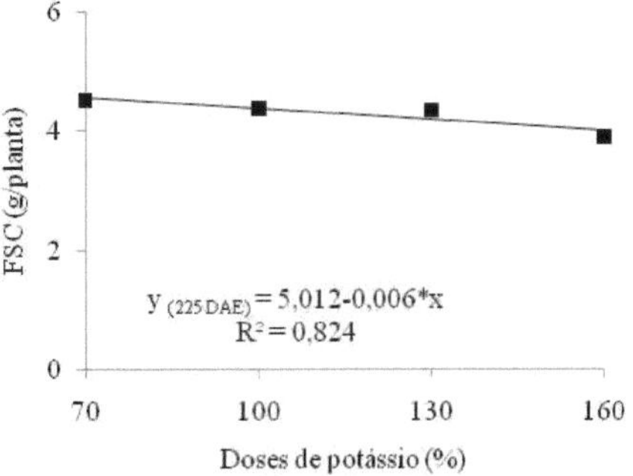

Figura 4. Stem dry phytomass - SCF of guava rootstocks cv. Paluma as a function of potassium doses at 225 DAE.

According to the summary of the analysis of variance (Table 3), there was a significant effect of the saline levels of the irrigation water on the total dry phytomass (p ≤0.01), root-ground ratio (p<0.05), leaf area ratio (p ≤0.05) and the Dickson quality index (p ≤0.01). However, the doses of potassium and the interaction between the factors (S x DK) did not significantly influence any of the variables studied.

Table 3. Summary of the analysis of variance for total phytomass (FST), root-air ratio (R/PA), leaf area ratio (RAF) and Dickson quality index (IQD) of guava rootstock cv. Paluma at 225 DAE as a function of different levels of irrigation water salinity and doses of potassium.

Source of variation	GL	Mean Squares			
		FST	R/PA	RAF	IQD
Salt levels (S)	4	$40,19^{**}$	$0,004^{*}$	$955,33^{*}$	$0,14^{**}$
Linear Reg.	1	$130,64^{**}$	$0,01^{*}$	$102,30^{ns}$	$0,54^{**}$
Quadratic rule	1	$20,58^{*}$	$0,00^{ns}$	$3333,03^{**}$	$0,00^{ns}$
Doses of K (DK)	3	$6,48^{ns}$	$0,01^{ns}$	$43,23^{ns}$	$0,01^{ns}$
Linear Regulation	1	$12,32^{ns}$	$0,01^{ns}$	$126,41^{ns}$	$0,02^{ns}$
Quadratic rule	1	$0,57^{ns}$	$0,00^{ns}$	$1,65^{ns}$	$0,00^{ns}$

Interaction (S*DK)	12	2.09^{ns}	0.01^{ns}	271.36^{ns}	0.00^{ns}
Blocks	3	10.47^{*}	0.02^{**}	642.90^{ns}	0.03^{ns}
CV (%)		13.58	15.34	23.55	14.63

ns, **, * respectively not significant, significant at p $\leq$ 0.01 and p $\leq$ 0.05

The salinity of the irrigation water negatively affected the total dry phytomass of the Paluma guava rootstocks at 225 DAE. The linear model (Figure 5A) indicated decreases of 10.88% per unit increase in the electrical conductivity of the irrigation water, i.e. a reduction in FST of 34.81% (3.5 g per plant) in plants irrigated with water of 3.5 dS m^{-1} , compared to plants irrigated with water with a lower ECa = 0.3 dS m^{-1} . Similar effects on the FST of guava rootstocks of the Ogawa and Paluma cultivars irrigated with water of different salinities were obtained by Gurgel et al. (2007) and Abrantes (2015) respectively.

The accumulation of $Na+$ and $Cl-$ ions in the leaves causes necrosis in the leaf tissues and accelerates the senescence of mature leaves with a consequent reduction in the photosynthetic area (MUNNS et al., 2006), 2006), so there will be less absorption of water and nutrients, consequently a reduction in plant growth and biomass accumulation (SOUTO et al. 2013), corroborating Sousa et al. (2011) and Sà et al. (2016), in the production of cashew and guava rootstocks, respectively.

At 225 DAE, the salt increase affected the root part area ratio of the guava rootstocks cv. Paluma and, according to the regression equation (Figure 5B), there was a linear decrease in the order of 4.78% per unit increase in ECa, i.e. a reduction of 15.30% (0.03 g g^{-1}) in the R/PA of plants irrigated with water of 3.5 dS m^{-1} in relation to those subjected to ECa = 0.3 dS m^{-1} , denoting the sensitivity of the crop. This reduction in the root-to-shoot ratio with increasing salinity was due to the higher rate of reduction in root phytomass compared to shoot phytomass, results which are in line with those obtained by Souza et al. (2016), when studying the effects of irrigation water salinity on the phytomass production of "Crioula" guava rootstocks; this difference between the parts of the plant may be related to the plant's physiological mechanisms for overcoming excess salts (NEVES, 2005) and/or to the characteristic difference of each species and/or genotype (AYERS & WESTCOT, 1999).

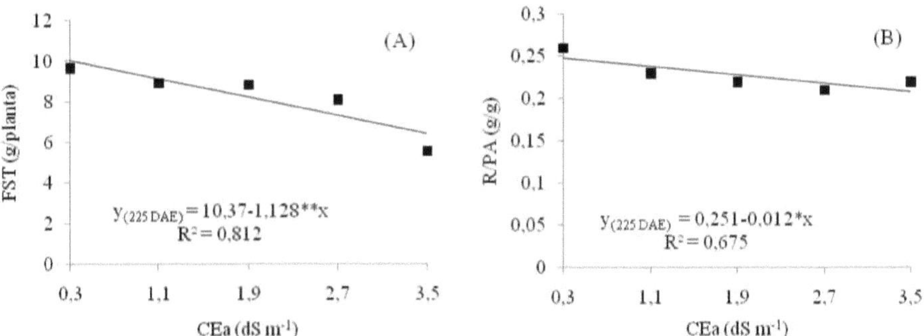

Figure 5 - Total dry phytomass - TDS (A) and root/ shoot ratio - R/PA (B) of guava rootstocks cv. Paluma as a function of the electrical conductivity of the irrigation water - ECa at 225 DAE.

With regard to the guava leaf area ratio (Figure 6A), it can be seen that the data fitted the quadratic model, with increasing RAF values being observed for the guava rootstocks cv. Paluma at 225 DAE up to the ECa level of 2.0 dS m^{-1} , where the maximum RAF value was 50.08 cm^2 g^{-1} . The leaf area ratio represents the leaf area per unit of mass produced in the plant, indicating the efficiency of dry matter formation by the photosynthetically active leaf area (MAGALHAES et al., 2007). In this sense, it can be seen that water salinity compromised the efficiency of the leaves in forming organic compounds.

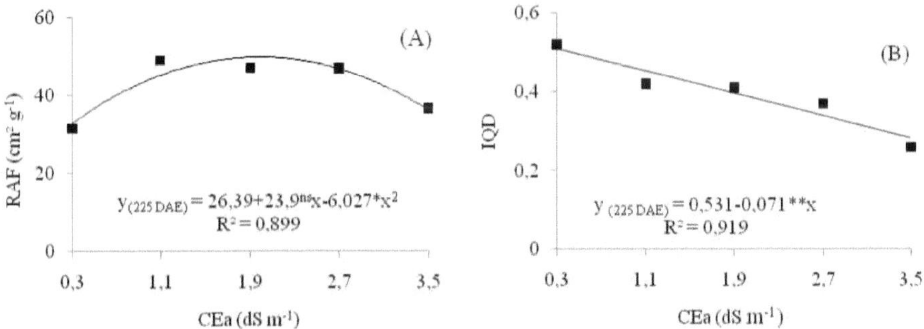

Figure 6: Leaf area ratio - RAF (A) and Dicson quality index - IQD (B) of guava rootstocks cv. Paluma as a function of the electrical conductivity of the irrigation water - CEa at 225 DAE.

The Dickson quality index (DQI) of guava rootstocks cv. Paluma at 225 DAE (Figure

6B), decreased as a function of water salinity with decreases of (13.37%) per unit increase in ECa, with a reduction of 34.81% (0.17) in plants irrigated with water of 3.5 dS m^{-1} in relation to those subjected to ECa = 0.3 dS m^{-1} . Similar data was obtained by Souza et al. (2016), when studying the effects of water salinity on the quality of guava rootstocks cv. Crioula, they obtained a decrease (12.24%) per unit increase and (39.16%) in plants irrigated with water of 3.5 dS m^{-1} in relation to those subjected to ECa = 0.3 dS m^{-1} .

Considering that the higher the value of the Dickson Quality Index (DQI), the better the quality of the seedlings (GOMES et al., 2003), as it is linked to the robustness and balance of the distribution of biomass by the correlation of various morphological characteristics in species such as guava (SOUZA et al., 2016), 2016). In this study, it was found that better quality guava rootstocks are obtained when the seedlings are irrigated with water with a lower saline content, since salinity impairs plant growth by altering the nutritional balance, cell metabolism, physiological and biochemical processes (HASANUZZAMAN et al., 2014).

4. CONCLUSIONS

1.   Irrigation with water of ECa 1.9 dS m^{-1} promotes an acceptable 10% reduction in phytomass production and quality of guava rootstocks cv. Paluma;

2.   The dose of 508.2 mg of K dm^{-3} of substrate favors the accumulation of dry phytomass in the stem of guava cv. Paluma at 225 DAE;

3.   There was no significant interaction (salt x K doses) on the variables studied.

5. BIBLIOGRAPHICAL REFERENCES

ABRANTES, D. S. **Interaction between salinized water and nitrogen fertilization in the production of grafted guava seedlings.** 2015. 91 f. Dissertation (Master's Degree in Agroindustrial Systems) - Federal University of Campina Grande, Pombal, 2015.

ALCARDE, C. A. Fertilizers. In: NOVAES, R. F.; ALVAREZ V, V. H.; BARROS, N. F.; FONTES, R. L. F.; CANTARUTTI, R. B.; NEVES, J. C. L. **Fertilidade do solo.** 1 ed., Viçosa: SBCS, 2007. p. 737-768.

AYERS, R. S.; WESTCOT, D. W. 1999. **Water quality in agriculture.** In: GHEYI, H. R.; MEDEIROS, J. L.; DAMASCENO, F. A. V. (Trad.). Campina Grande: Federal University of Paraiba. 153 p. (FAO Studies: Irrigation and Drainage, 29 Revised).

BENINCASA, M. M. P. **Plant growth analysis, basics.** 2 ed. Jaboticabal: FUNEP, 2003. 41p.

BERNARDO, S.; SOARES, A. A.; MANTOVANI, E. C. **Manual de irrigação.** 8 ed., Viçosa: UFV, 2006. 625 p.

BRAZIL. Ministry of Agriculture. Research and Experimentation Office. Pedology and Soil Fertility Team. **I Exploratory and reconnaissance survey of soils in the state of Paraiba. II Interpretação para uso Agricola dos Solos do Estado da Paraiba.** Rio de Janeiro: 1972. 638p. (Technical Bulletin, 15; SUDENE, Pedological Series, 8).

BRITO, M. E. B.; FERNANDES, P.D.; GHEYI, H.R.; MELO, A.S.; SOARES FILHO, W.S.; SANTOS, R.T. Sensitivity to salinity of trifoliate hybrids and other citrus rootstocks. **Revista Caatinga**, Mossoró, v. 27, n. 1, p. 17-27, 2014.

CAVALCANTE, L. F; VIEIRA, M. S.; SANTOS, A. F.; OLIVEIRA, W. M.; NASCIMENTO, J. A. M. Saline water and liquid bovine manure in the formation of seedlings of guava cultivar Paluma. **Revista Brasileira de Fruticultura**, Jaboticabal, v. 32, n. 01, p. 251-261, 2010.

CLAESSEN, M. E. C. (org.). **Manual of soil analysis methods**. 2. ed. rev. atual. Rio de Janeiro: Embrapa-CNPS, 1997. 212p. (Embrapa-CNPS. Documentos, 1).

CORRÊA, M. C. M.; PRADO, R. M.; NATALE, W.; PEREIRA, L.; BARBOSA, J. C. Responses of guava seedlings to doses and modes of application of phosphate fertilizer. **Revista Brasileira de Fruticultura**, Jaboticabal, v. 25, n. 1, p. 164-169, 2003.

DIAS, M. J. T.; SOUZA, H. A.; NATALE, W.; MODESTO, V. C.; ROZANE, D. E. Nitrogen and potassium fertilization for guava seedlings in a commercial nursery. **Semina: Ciências Agrárias**, Londrina, v. 33, supplement 1, p. 2837-2848, 2012.

DICKSON, A.; LEAF, A. L.; HOSNER, J. F. Quality appraisal of white spruce and white pine seedling stock in nurseries.**The Forest Chronicle**, v. 36, n. 01, p. 10-13, 1960.

BRAZILIAN AGRICULTURAL RESEARCH COMPANY - EMBRAPA. National Soil Research Center. **Brazilian Soil Classification System**. 3. ed. Brasilia, 2013. 353 p.

EPSTEIN, E.; BLOOM, A. J. **Nutriçâo mineral de plantas:** principios e perspectivas. 2. ed. Londrina: Editora Planta, 2006. p. 169-236.

FACHiNELLO, J. C.; NACHTiGAL, J.C.; HOFFMANN, A. Propagation by seeds. iN: FACHiNELLO, J. C.; HOFFMANN, A.; NACHTiGAL, J.C. (Ed.). **Propagation of fruit plants**. Brasilia, DF: Embrapa Informaçao Tecnològica. 2005. P. 57-67.

FERREiRA, D. F. SiSVAR: A computer statistical analysis system. **Ciência e Agrotecnologia**, v. 35, n. 06, p. 1039-1042, 2011.

FRANCO, F. C.; PRADO, R. M.; BRACHiROLLi, L. F.; ROZANE, D. E. Growth curve and macronutrient uptake in guava seedlings. **Revista Brasileira de Ciência do Solo**, Viçosa, v. 31, n. 6, p. 1429-1437, 2007.

GOMES, J. M.; COUTO, L.; LEiTE, H. G.; XAViER, A; GARCiA, S. L. R. Growth of Eucalyptusgrandis seedlings in different sizes of tubes and N-P-K fertilization. **Revista Arvore**, Viçosa, v. 27, p. 113-127, 2003.

GURGEL, M. T.; GHEYI, H. R.; FERNANDES, P. D.; SANTOS, F. J. S.; NOBRE, R. G. Initial growth of guava rootstocks irrigated with saline water.**Revista Caatinga**, Mossoró, v.20, n.2, p.24-31, 2007.

HASANUZZAMAN M., ALAM M. M., RAHMAN A., HASANUZZAMAN M., NAHAR K., FUJITA M. Exogenous proline and glycine betaine mediated upregulation of antioxidant defense and glyoxalase systems provides better protection against saltinduced oxidative stress in two rice (Oryza sativa L.) varieties. **BioMed Research Internatinal**, Juazeiro do Norte, v. 1, p. 1-17, 2014. Available from: <http://dx.doi.org/10.1155/2014/757219>

IBGE - Brazilian Institute of Geography and Statistics (2016). Sidra - **Municipal Agricultural Production**, 2016. Available at:< http:// www.sidra.ibge.gov.br> (Accessed December 12, 2016).

LIMA, L. G. S.; ANDRADE, A. C.; SILVA, R. T. L.; FRONZA, D.; NISHIJIMA, T. **Mathematical models for estimating the leaf area of *guava* trees (*PsidiumguajavaL.*).** In: 64ª SBPC ANNUAL MEETING. Sao Luiz: UFMA, 2012.

MAGALHÂES, P. C.; DURÂES F. O. M.; RODRIGUES, J. A. S. **Ecophysiology**. In: RODRIGUES, J. A. S. (Ed.). Cultivation of sorghum. 5. ed. Sete Lagoas: Embrapa Milho e Sorgo, Embrapa Milho e Sorgo. Sistemas de produção, 2007. 30p.

MALAVOLTA, E. Potassium: absorption, transport and redistribution in the plant. In: YAMADA, T.; ROBERTS, T. L. (ed.). **Potassium in Brazilian agriculture**. Piracicaba: Brazilian Association for Potash and Phosphate Research, 2005. chap. 8, p. 179-238.

MEDEIROS, J. F. DE; LISBOA, R. A.; OLIVEIRA, M.; SILVA JÛNIOR, M.J.; ALVES, L. P. Characterization of groundwater used for irrigation in the melon-producing area of Chapada do Apodi. **Revista Brasileira Engenharia Agricola e Ambiental,** Campina Grande, v. 7, n.3, p. 469- 472, 2003.

MEDEIROS, P. R.; DUARTE, S. N.; UYEDA, C. A.; SILVA, E. F. F.; MEDEIROS DE, J. F. Tolerance of the tomato crop to soil salinity in a protected environment.

Revista Brasileira Engenharia Agricola e Ambiental, Campina Grande, v. 16, n. 01, p. 51-55, 2012.

MUNNS, R.; JAMES, R.A.; LÄUCHLI, A. Approaches to increasing the salt tolerance of wheat and other cereals.**Journal of Experimental Botany,** Oxford, v.57, p.1025-1043, 2006.

NEVES, G. Y. S. et al. Seed germination and seedlings growth of soybean (*Glycine max* (L.)Merr.)under salt stress. **Bioscience Journal**, Uberlândia, v. 21, p. 77-83, 2005.

PEREIRA, F. M.; CARVALHO, C.; NACHTIGAL, J.C. Século XXI: a new dual-purpose guava cultivar. **Revista Brasileira de Fruticultura**, Jaboticabal, v. 25, n.3, p.498-500, 2003.

RICHARDS, L. A. (ed.) **Diagnosis and improvement of saline and alkali soils**. Washington, United States Salinity Laboratory: 1954. 160p. (USDA. AgricultureHandbook, 60).

RAMOS, D. P.; SILVA, A. C.; LEONEL, S.; COSTA, S. M.; DAMATTO JÙNIOR, E. R. Production and fruit quality of the 'Paluma' guava tree, subjected to different pruning times in a subtropical climate. **Revista Ceres**, Viçosa, v. 57, n.5, p. 659-664, 2010.

SA,F. V. S.; BRITO, M. E. B.; FERREIRA, I. B.; ANTONIO NETO , P.; SILVA, L. A.DE; COSTA DA, F. B. Salt balance and initial growth of pine seedlings (*Annonasquamosa* L.) under substrates irrigated with saline water. **Irriga**, Botucatu, v. 20, n. 3, p. 544-556, 2015.

SA, F. V. S; NOBRE, R. G.; SILVA, L. A.; MOREIRA, C. L.; PAIVA, E. P.; OLIVEIRA, F. A. Tolerance of guava rootstocks to salt stress. **Revista Brasileira de Engenharia Agricola e Ambiental**, Campina Grande, v.20, n.12, p.1072-1077, 2016.

SILVA, E. M.; NOBRE, R. G.; GHEYEI, H. R.; PINHEIROS, F. W. A.; SOUZA, L.

de P.; DIAS, A. S. **Growth and quality of guava rootstock under irrigation with salinized water and nitrogen doses.** In: III Inovagri International Meeting, Fortaleza, CE. 2015.

SILVA, S. M. S.; ALVES, A. N.; GHEYI, H. R. Development and production of two cultivars of papaya under saline stress, **Revista Brasileira de Engenharia Agricola e Ambiental**, Campina Grande,v.12, n.4, p.335-342, 2008.

SOUSA, A. B. O.; BEZERRA, M. A.; FARIAS, F. C. Germination and initial development of common cashew clones under irrigation with saline water. **Revista Brasileira de Engenharia Agricola e Ambiental**, Campina Grande, v.15, p.390-394, 2011.

SOUTO, A. G. L.; CAVALCANTE, L. F.; NASCIMENTO, J. A. M.; MESQUITA, F. O.; LIMA NETO, A. J. Behavior of noni to irrigation water salinity in soil with bovine biofertilizer. **Irriga**, Botucatu, v.18, p.442- 453, 2013.

SOUZA, L. P.; NOBRE R. G., SILVA E. M.; LIMA, de G. S.; PINHEIRO, F. W. A.; ALMEIDA L. L. de S. Formationof 'Crioula' guavarootstockunder saline Waterirrigationandnitrogen doses. **Revista Brasileira de Engenharia Agricola e Ambiental,** Campina Grande, v.20, n.8, p.739-745. 2016.

Printed by Books on Demand GmbH, Norderstedt / Germany